SEASHELLS

— Jewels from the Ocean —

Budd Titlow

Voyageur Press

DEDICATION

To my dad, Franklin H. Titlow Jr.; my mom, Chesnut "Ches" Budd Gearing; my three wonderful children—daughters Mariah Titlow and Merisa Titlow and son David Denny; and, most importantly, to my beautiful wife, Deborah "Debby" Igleheart Titlow.

ACKNOWLEDGMENTS

Dad's 25 years of shelling Florida's coasts along with his lovely wife, Elizabeth "Liz," gave me the inspiration and much of the material—shells and writing—for this book. Mom and her husband, Dr. Frank W. Gearing Jr., provided continuous "you can do it" encouragement. Mariah, Merisa, and David also provided critical voices of support and many well-timed edits throughout this two-year project. Finally, Debby lovingly tolerated my weekend after weekend disappearances into my basement office and lent her substantial artistic talents to composing many of the book's photos.

   I also owe a debt of deep gratitude to the following people who provided shells, photos, and other support for the creation of this book:

- Aunt Betty Igleheart, Uncle Elliott "Ike" Igleheart, and Cousin Lucie Johnson, all beloved relatives and dear friends
- My sister, Meg James, and her family, Kent, Ryan, Ashley, and Mary Kate
- Kevin Adams, North Carolina's best nature photographer
- Shawsheen Baker, landscape architect extraordinaire
- Paul Callomon, technical review and editing

First published in 2007 by Voyageur Press, an imprint of MBI Publishing Company LLC, Galtier Plaza, Suite 200, 380 Jackson Street, St. Paul, MN 55101 USA

Editor: Leah Noel
Technical Reviewer: Paul Callomon
Designer: Jennifer Bergstrom

Printed in China

The information in this book is true and complete to the best of our knowledge. All recommendations are made without any guarantee on the part of the author or Publisher, who also disclaim any liability incurred in connection with the use of this data or specific details.

Voyageur Press titles are also available at discounts in bulk quantity for industrial or sales-promotional use. For details write to Special Sales Manager at MBI Publishing Company, Galtier Plaza, Suite 200, 380 Jackson Street, St. Paul, MN 55101 USA.

To find out more about our books, join us online at www.VoyageurPress.com.

Library of Congress Cataloging-in-Publication Data

Titlow, Budd.
   Seashells:jewels from the ocean/by Budd Titlow.
      p. cm.
   Includes bibliographical references and index.
   ISBN: 978-0-7603-2593-3 (plc w/ jacket)
   1. Shells. 2. Mollusks. I. Title.
QL403.T58 2007
594.147'7--dc22
                                   2007021891

On the cover: A chambered nautilus (*Nautilus pompilius*), a unique member of the cephalopod class of mollusks. *Shutterstock/Suzanne Tucker*

On the front flap: A kaleidoscope of color from the natural world of mollusks.

On the back cover: main: A beautiful beach scene with a gastropod shell; **insets:** a common American Sundial (*Architectonica nobilis*), an alphabet cone (*Conus spurious*), and a pear whelk (*Busycon spiratum*), gastropods.

Contents

Preface

I grew up in the mountains, but my heart has always yearned for the sea. As a youngster, this meant cajoling as many trips as I could muster to my dad's house on Sandbridge Beach in Virginia. The house was quite small—not much bigger than an apartment—and not fancy at all. Dad never quite finished the inside, so the walls consisted of 2x4-foot studs covered with tar paper. The floors never advanced past the wooden underplanking stage, and showering there involved a cold-water device mounted to the outside of the house. But for a down-and-dirty nature rat like me, Dad's sand-dune contrivance was exactly what a beach house should be. Not needing to worry about damaging the baubles and bangles of the mansions lining the beach, I was free to use the house as a combination clubhouse/research station, often dissecting curious sea creatures right in the middle of the porch floor.

My Sandbridge days were spent dangling rotten chicken necks in front of blue crabs, braving head-high waves to sling blood worms into the surf and, when the moon was right, catching buckets of croakers and spot. Generally considered trash fish by local sportsmen, these small, surf-schooling fish were all trophies to me.

Unable to sit still on a beach for more than about a second, I always hated the lounge chair, beach umbrella, and sandcastle-building routines. Instead, I took long walks along the wrack line at the top of the intertidal zone, dashing and splashing through each breaking wave, loving the squish of wet sand squirting up through my toes, and stopping to examine everything that caught my eye.

It was during just such a beachcombing venture in the summer of 1956 that I had my first up-close-and-personal encounter with mollusks, although I certainly didn't know them by that name at the time. I noticed that each receding wave unearthed a palette of miniature jewel-like shells, all furiously digging in unison back down into the wet sand.

A call to my dad brought a typical response: "Let me show you something really fun to do!" With that, he marched down into the breakers, turned his backside to the ocean, and screwed his feet down into the

Imagine being able to walk into this picture and you'll understand the allure of shell collecting. *Shutterstock/Emin Kuliyev*

Each receding wave revealed a palette of miniature jewel-like Atlantic coquina shells *(Donax variabilis)*. *Kevin Adams*

The first time I dug down into the wet sand and pulled up two handfuls of tiny jewels, I was hooked. *Shutterstock/Anastasiya Igolkina*

sand. As soon as the next wave passed by and began to retreat, he dug his hands deep into the sand and pulled up hundreds of tiny, multicolored shells.

Now I was really hooked! I rushed back to the beach house, grabbed my plastic bucket, and sped back to the beach to collect some new pets. I just couldn't believe that this soggy, seemingly sterile sand was hiding hundreds of tiny jewels—each one with its own distinctly different colors and patterns. With each double handful of sand, I was unearthing a new gallery of nature's art.

The shells I was finding were tiny bivalves known as Atlantic coquinas *(Donax variabilis)*. They were filter feeding in the surf, being uncovered by outgoing waves, and using their tiny "feet" to dig back down before the next set of waves exploded across the intertidal zone.

I figured that if I kept 15 to 20 of the prettiest shells each day, by the end of my two weeks at the beach I would have a collection that would be the envy of my friends back in my mountain home in Christiansburg, Virginia. Of course, my plan didn't work out very well. Without the revitalizing nourishment of the tidal action, my prized works of art quickly overheated and died in my plastic bucket. I realized this only after my third day of collecting, when my bucket gallery started to stink like the inside of a barnyard and I was summarily instructed to take the whole smelly mess outside and bury it.

Twenty-five years later, when I returned to Sandbridge Beach from my home in Colorado, I couldn't wait to share the thrill of finding jewels in the sand with my two young daughters. Buckets in hand, we all took our requisite digging positions. Right on cue as a wave rolled in and out, we all pulled up soupy handfuls of wet sand and, to the chagrin of my excited daughters, not much else. Another wave rolled in, and again we all dug down deep and pulled up nothing. We changed positions,

moving first several hundred feet up the beach and then several hundred feet down the beach.

There was not a coquina to be found on Sandbridge Beach that day, or during the rest of our visit that year, or during our visits since. Apparently, something had caused the local population of coquinas to go from millions to seemingly nonexistent in a mere 25 years. Also, Sandbridge's resident crustacean community, notably sand fleas and ghost crabs, were similarly missing in action.

I soon learned from my dad that this catastrophic loss of beach life was not due to just one "something" but actually a long, interrelated string of poor environmental management decisions. The series of mistakes began when the federal government approved construction of massive bulkheads to protect Sandbridge's oceanfront mansions from winter's nor'easters. This led to the loss of the intertidal zone, sending high tide waves smashing against the bulkheads and leaving only a 25-foot-wide strip of beach at low tide. Of course, it took only two winters for the ocean's surging storm power to rip out the bulkheads and reclaim some of its natural tidal range.

Unfortunately, the bulkhead fiasco permanently

A father teaches his young daughter the finer arts of shell collecting on Sanibel Island in Florida, the best place to hunt for seashells in the United States.

Many beaches now restrict collection of live mollusks. Always check the local regulations before you begin collecting!

changed Sandbridge's coastal geology and associated sediment flow dynamics. The resulting shift in beach erosion and deposition practically eliminated Sandbridge's once wide and wonderful beaches. Now each spring the government has to pump in millions of cubic yards of dredged sand to restore the depleted Sandbridge beaches for the coming summer recreation season.

In April 2006, 50 years after my official introduction to mollusks at Sandbridge Beach, I arrived at Sanibel Island, Florida, the proclaimed shelling capital of North America. I traveled there to conduct firsthand research for this book. My primary goal was to personally experience the world of big-time shell collecting, more technically known as conchology. I also wanted to get some nice shots of ardent collectors pursuing their quarries and visit the wonderful Bailey-Matthews Shell Museum, one of the most comprehensive repositories of information on shells and their relation to human populations throughout history.

My first sunrise on Sanibel turned out to be quite a revelation. By all accounts, it was just a normal day at the beach. There had not been any recent storms that would generate currents strong enough to rip

whelks, conchs, tulips, and the like from the ocean bottom and deposit them to die and be collected on Sanibel's shores.

As I stood in predawn light, expecting to be completely alone with the sunrise, all I could see in both directions were people, all demonstrating the Sanibel Stoop, the distinctive posture that shouts "shell collector" to anyone who sees it. Worst of all, I didn't see any newly arrived intact shells anywhere on the beach. So much for my plans for shooting a Sanibel sunrise with a foreground full of beautifully colored and shaped shells.

From ensuing conversations with Sanibel Islanders and visiting collectors, I learned that my first day's experience was par for the course. Generally speaking, you seldom find intact shells on Sanibel's beaches anymore. They all get collected, literally within minutes of rolling in from the surf.

I also learned that on good shelling days when an overnight storm deposits a wealth of new shells on Sanibel's beaches, the competition really turns fierce. As soon as the storm ends, droves of collectors show up with flashlights and plod through the still rough surf in complete darkness—taking the best pickings before the sun comes up. As a result of the intense, highly competitive emphasis now placed on collecting, many of Sanibel's once-common shells are now difficult to find.

As a lead-in to this book, I've told you about two of my most memorable experiences with mollusks. Separated in time by exactly 50 years, they both began as exhilarating natural adventures and ended disappointingly with the disappearance of the animals that attracted me in the first place.

Yet my intention here is *not* to put a negative spin on the information that you are about to read. In fact, you'll find that 85 percent of my book—the first seven chapters—is an *Ode to Joy* of the molluscan world. In these pages, you'll explore the intricacies, foibles, facts, and fantasies that go beyond the known world of seashells and deep into the mostly unknown world of their living counterparts: mollusks.

Most of us have heard the "roar of the ocean" in a large seashell. *Shutterstock/Stephen Coburn*

Then, near the end of the book, in Chapter Eight, you'll revisit the experiences I just described and ask the question that must be answered: Are we mistreating, and simultaneously loving, our mollusk populations and other coastal resources to death?

This book is primarily intended to increase readers' awareness of the beauty, uniqueness, value, and enjoyment of mollusks. But my secondary goal is to build a bandwagon of support for mollusks and the other coastal creatures that share the planet with us. If you think this book accomplishes these goals, please let me know. I'd love to hear from you!

Budd Titlow
3420 Fairway Lane
Durham, North Carolina 27712
btitlow@aol.com

Children need only these basic collecting tools to go seashell hunting and entertain themselves for hours. *Shutterstock/Lorraine Kourafas*

Introduction

For a child, the natural world is full of wondrous marvels—fireflies, shooting stars, and rainbows, to name just a few. But perhaps most mysterious and awe-inspiring of all are seashells. Who doesn't remember being enchanted by seashells on a first trip to the beach—looking in fascination at the incredible variety of shapes, colors, and patterns . . . listening in wonder to the roar of the surf, somehow trapped inside a shell's inner sanctum? Wondering what created this spectacular gallery of natural art, which seems to change completely with each new high tide. What created these masterpieces? How did they get here?

As you grew older, you learned that seashells are the remains of a remarkable group of organisms called mollusks. As one of the most abundant organisms in the world's oceans, mollusks are critical to sustaining life in the sea. Alive, they form one of the broad bottom layers of the ocean's food pyramid, sustaining the pyramid's upper layers—everything from spiny lobsters to giant bluefin tuna. After they die, mollusks' calcium-rich shells are used for a variety of needs throughout the oceanic community. Many become pre-owned homes for such bottom-dwelling nomads as the hermit crab. Others are broken down and crushed by the ocean's energy, over time supplying building materials for large-scale marine housing projects such as coral reefs.

Most of us, however, don't realize the incredibly significant role mollusks and seashells have played in the earth's human history. Throughout the world's civilizations, mollusks and their shell homes have been used for everything from food supplies to cooking tools, currency, and communication. The ancient Greeks used giant conch shells to communicate from mountaintop to mountaintop. The Calusa Indians fashioned deadly weapons from whelk shells and used them to defend their south Florida homeland from invading Spaniards. And then in 1626, Dutch colonists concocted an infamous scheme to buy Manhattan Island from the Delaware Indians for $24 worth of trade goods, including beads made from seashells. Now think what Manhattan is worth—as they say in Brooklyn, "Now dat's a lotta clams!" Even today, mollusks are the number one food source harvested from the world's oceans.

For centuries, both architects and artists have been inspired by the deep, spiraling symmetry on the underside of this Common American Sundial (*Architectonica nobilis*).

An array of colorful seashells shares the beach with sand dollars and starfish on Sanibel Island, Florida.

This book will immerse you in the bewildering diversity and fascinating human and natural history of seashells. It will discuss the most famous shelling hot spots on the Gulf, Atlantic, and Pacific coasts of North America.

This book will also explore the sizes, shapes, habits, and life histories of the mollusks that occupy some of the world's most exotic and beautiful shells, as well as examine how mollusks grow shells and how they withstand the powerful currents on the ocean bottom.

Next, you'll learn how and why these colorful creatures became so important to human populations. You'll even discover how shells are closely tied to North American folklore, major historical events, religious movements, myths, legends, and pop culture.

Finally, this book will look at the primary hazards to mollusk survival—commercial fishing, poor land development practices, and habitat destruction—and what we all can do to lessen these threats.

Among nature's most perfect creations, seashells leave their beauty with us for generations—sometimes for eons as fossils—after they die. *Shutterstock*

Chapter 1

What's a Mollusk?
AND WHICH CAME FIRST: THE MOLLUSK OR THE SHELL?

Seashells are ubiquitous, one of Mother Nature's most common and endearing creations. But these well-known and well-loved objects are formed by one of the most abundant but least understood animals: mollusks. Most people know what a seashell is, but mentioning the word "mollusk" often leads to quizzical looks and shaking heads. And if the question of how seashells are created is posed, the likelihood of the response including words like "mantle" and "calcium carbonate" is remote. In truth, there is a worldwide disconnect between the sight and the source of one of our most abundant and important natural resources.

If it weren't for mollusks, there would be no seashells. Mollusks are inside-out organisms. They literally grow seashells to support and protect their soft, fleshy bodies (see Chapter Four for details). When mollusks die, their soft bodies rapidly decompose, leaving their hard, outer shells behind for the world to find and collect.

The phylum Mollusca is the second-largest group of animals in the world, with over 100,000 species in it. The phylum Arthropoda, which includes the world's insects, far and away ranks number one in abundance, but when it comes to diversity in size, shape, and appearance, mollusks are the champs.

Mollusks are invertebrates, meaning they have no internal backbones. They grow shells to literally hold themselves together. Scientifically, seashells are the exoskeletons, or outer frames, of mollusks. The shells you see on the beach are the remains of dead mollusks, much like the bones of cattle bleaching in the blast furnace sunlight of California's Death Valley desert.

While all seashells are produced by mollusks, not all mollusks produce shells. In fact, the most exciting mollusks—those featured in Hollywood movies— are shell-less. Octopuses have lost their shells altogether, while squid have mostly organic, chitinous internal shells. These and other cephalopods will be discussed in Chapter Two.

NO BEACH FUN FOR MOLLUSKS

For humans, a day at the beach is one of life's greatest rewards. Nothing on earth compares with a refreshing, relaxing trip to the ocean's shore. But for mollusks,

The sea urchin (top) seems to always hang around with seashells, but it does so only because it's hungry—not because it likes being with kin. Sea urchins are not mollusks, but echinoderms.

Mythical Monsters of the Deep

Fictional accounts of sea monsters resembling giant squid have fascinated readers since the time of Homer. In his novel, *Toilers of the Sea*, Victor Hugo portrayed giant human-eating squid. Norwegian sailors have their legend of the Kraken, in which a tentacled sea monster as large as an island is capable of engulfing and sinking any ship. Stories of swashbuckling bravado with sea monsters continued in books such as Herman Melville's *Moby Dick* and Jules Verne's *20,000 Leagues Under the Sea*.

The giant squid (family Architeuthidae) featured in all these works are real, and they really can reach mythical proportions in the natural world. Size estimates range up to over 50 feet in length and 1,980 pounds in weight. In the 2005 book, *Out of My Shell*, S. Peter Dance provides an intriguing, albeit unconfirmed, account that suggests giant squid may grow even larger.

According to Dance, one night during World War II, a British trawler was lying off the Maldive Islands in the Indian Ocean. While fishing alone on the deck, seaman A. G. Starkey saw something in the water. He described it by noting the following:

> As I gazed, fascinated, a circle of green light glowed in my area of illumination. This green unwinking orb I suddenly realized was an eye. The surface of the water undulated with some strange disturbance. Gradually, I realized that I was gazing almost point-blank range at a huge squid.

Seaman Starkey then claimed to have walked the length of the ship, finding the squid's tail at one end and the tentacles at the other. The ship was 175 feet long.

Whether Starkey's tale can be believed is somewhat of a moot point. Giant squid are, without question, the world's largest invertebrates. Giant squid also possess the largest eyes, reaching diameters of 18 inches, of any living creature. If Starkey did happen to be a little tipsy that night on the high seas, the mere sight of an 18-inch eye must have certainly shocked him back into sobriety.

Giant squid are very well adapted to living as aggressive predators. Their arms are equipped with hundreds of 1- to 2-inch-wide suction cups, each mounted on an individual stalk and encircled with a ring of sharp teeth. These suction cups aid the creature in capturing its prey by firmly holding it by both suction and perforations. These monsters of the deep also have beaklike mouths that are strong enough to cut through steel cable.

Until September 30, 2004, no one on earth had ever seen a giant squid alive in its natural habitat, the pitch-dark waters of ocean trenches up to 3,300 feet deep. Then a duo of Japanese scientists succeeded in rigging what amounted to the world's longest trotline (3,000 feet), hung with tantalizing bait balls of squid and shrimp topped off with a camera and several strobe lights.

After about 20 passes through known sperm whale habitat, a 26-foot giant squid took the bait so viciously that it impaled a tentacle in the process. During the four hours it took the squid to break free, the intrepid scientists ended up with more than 500 photographs and one pre-owned squid tentacle.

the beach is often the beginning of the end. The high-tide line is the last place a mollusk wants to end up because it means that a disaster has struck and its homeland—often a coastal shelf covered by 30 to 100 feet of water—has been torn apart. A mollusk disaster usually involves a strong winter storm with ocean-roiling winds. The winds bite down deep into the ocean waters, creating currents that scour the sea bottoms and sweep everything that isn't securely anchored, including most mollusks, onto the shores of the nearest beaches. As the storm recedes, the mollusks are left high, dry, and at the mercy of either drying up or being preyed upon, both leading to death. Soon afterward, the live parts of the mollusks have vanished, leaving only their brightly colored outer shell coverings for delighted beachcombers and joyful children to find and enjoy.

Looking like they're posing for a ballet, these boldly whorled wentletraps (family Epitoniidae) surround an exquisitely colored cone shell (family Conidae). Exotic shells like these motivate serious conchologists to travel the world in search of new additions to their collections. *Shutterstock/Popovici Ioan*

Even a novice collector seldom has trouble identifying this shell. The shape, patterns, and colors of this bivalve perfectly mimic its namesake: turkey wing (*Arca zebra*).

Mussel shells accentuate a natural bouquet of flowers on Monhegan Island, Maine.

Barnacles, snails, a mussel, and a limpet inhabit the bottom of a crystal-clear tidepool in Two Lights State Park on Cape Elizabeth, Maine.

BEAUTY FROM THE BEASTS

Many of earth's amazing creatures begin life in ordinary packaging and become more beautiful and unique as they mature. Nondescript eggs yield exquisitely colored wood warblers. Bland, brown caterpillars turn into brilliant butterflies. Boldly colored dragonflies emerge from wrinkled chrysalises. With these species, their "birth packages" innocuously disappear and return to dust. And at death, their remains are also visually unpleasant.

But with mollusks, Mother Nature throws a curve. Decidedly unattractive animals at birth, mollusks create unparalleled works of art as they grow. Then they leave their perfect creations—seashells—for us to enjoy long after their deaths.

RIGHT
In addition to the joy of finding perfect specimen shells, beachcombing often leads to other interesting discoveries—like this patch of seaweed that has formed a purple "seahorse."

BOTTOM
Five snails in the bottom of a deep tidepool in Reid State Park, Maine.

Rumors persist about giant human-eating clams. While clams do grow to 4.5 feet in length and 500 pounds in weight, their inability to open their mouths more than 6 inches prevents them from eating anything big—like people. In fact, clams are primarily filter feeders, gleaning nutrients from the plankton in the ocean. *Shutterstock/Steffen Foerster Photography*

MONET, MANET, AND MOLLUSKS

When painting landscapes on Monhegan Island in Maine, Jamie Wyeth—son of American realist painter Andrew Wyeth and grandson of renowned illustrator N. C. Wyeth—hid inside a cardboard box to escape public scrutiny. In much the same way, mollusks remain hidden to human eyes while creating their shell masterpieces. As a result, we know only the creations (seashells) and not the creators (mollusks).

Like the Wyeths, mollusks are among the world's most amazing artists. Their works are the masterpieces of oceanic museums. Their shapes, colors, textures, and patterns are endless in their variety. For eons, their shell works have inspired human artists, architects, and craftsmen.

A BEACH IS AN ART GALLERY IN FLUX

Life at the seashore is controlled by the tides. Literally everything that happens is built either into or around the tidal cycle. From an artistic perspective, the tides are the curators of shoreline galleries. Twice each day, high tides roll in, sweep the sand canvases clean, and deposit new exhibits. Seashells are the featured masterpieces of these constantly rotating exhibits of exquisite natural beauty.

A Case of Mistaken Identity

Beachcombers often find more than mollusks on their shell-hunting trips. While sea urchins, starfish, and sand dollars look like shell-bearing mollusks, they aren't. These fanciful collector's items are actually echinoderms.

Echinoderms are quite different from mollusks in the following ways:

- An echinoderm's body is covered by an epidermis that consists of an internal living basal layer and an external dead horny layer. (The majority of mollusks are protected by hard, calcareous outer shells.)
- All echinoderms are marine creatures; none live in freshwater or on land. (Thousands of molluscan species—primarily snails, slugs, and mussels—live on land or in freshwater.)
- Echinoderms possess radial symmetry. In fact, all echinoderms exhibit fivefold radial symmetry in portions of their body at some stage of life. (Mollusks either have bilateral symmetry—the bivalves—or no symmetry—the univalves or gastropods.)
- Echinoderms have simple nervous systems, but no brains. (Some mollusks have complex nervous systems, and all have brains.)
- An echinoderm's body consists of (usually five) equal segments, each containing a duplicate set of essential internal organs. Look closely at a sand dollar or a sea urchin. Even though they appear to be round single units, in reality they are nothing more than five-armed starfish with their legs wrapped inward to form a sphere. (Mollusks do not have duplicate sets of organs.)
- Many echinoderm species can regenerate themselves. A sea star that is cut radially into a number of parts will, over a period of months, regenerate into as many separate, viable sea stars. (Mollusks do not have regenerative powers.)
- Many echinoderms, such as sea stars, feed by extruding their stomachs out from their bodies. In a scene worthy of any horror movie, an extended stomach will then surround, devour, and digest a living prey species. (Mollusks cannot extrude their stomachs.)

Remarkably, these four creatures that collectors often see together represent different phyla in the animal kingdom. The lightning whelk is a mollusk, the sand dollar and starfish are echinoderms, and the coral represents the phylum Cnidaria.

Chapter 2

The Shell Game
The Classification and Types of Mollusks

While up to 100,000 species of mollusks are currently classified, new species are still being found and named every year. During 2004 and 2005, a French-led marine expedition team surveyed the waters around Panglao, a Philippine island lying 390 miles southeast of Manila. According to results released in February 2007, the survey team estimated that 1,500 to 2,000 species of the mollusks it found in the region were new species.

While mollusks are most often associated with the ocean, they can be found in nearly every ecosystem on earth, ranging from the lowland rainforests of the Amazon to the barren mountaintops of the Himalayan Range. The most common nonmarine mollusks are freshwater and terrestrial snails. They're found everywhere, from the tiniest creeks to backyard gardens. In fact, clams and mussels, normally considered coastal dwellers, are also found throughout the world's freshwater drainage basins.

Most of these freshwater species evolved from saltwater species that invaded freshwater river systems in coastal areas. Today, due to water pollution and loss of habitat, freshwater mollusks are among the most imperiled groups of animals. Thirteen species of freshwater mussels are protected under the U.S. Endangered Species Act, including the dwarf wedgemussel (*Alasmidonta heterodon*), Tar River spinymussel (*Elliptio steinstansana*), and the Carolina heelsplitter (*Lasmigona decorata*).

As some of the oldest creatures on earth, mollusks date back to the Precambrian Period. They first appeared in the fossil record some 500 million years ago, meaning that the first mollusks were swimming in the ancient seas some 300 million years before dinosaurs roamed the earth.

Mollusks range in size from microscopic clams to giant squid. Almost 1,000 species—including oysters, clams, tulip shells, angel wings, apple murexes, horse and fighting conchs, lions-paw scallops, junonias, and pear and lightning whelks—can be found on Florida's Outer Continental Shelf alone. The smallest mollusk, the *Pythina* clam, is the size of a grain of rice. Tiny, smooth, and translucent, this miniature bivalve attaches to the undersides of shrimp and crayfish. The largest known bivalve seashell, the giant clam (*Tridacna maxima*) of the

Throughout history, artists and craftsmen have been inspired by the scallop's elegant shape and vibrant colors. This is a calico scallop (*Argopecten gibbus*).

South Pacific, grows up to 4.5 feet long and can tip the scales at 500 pounds.

FIVE PRIMARY CLASSES OF MOLLUSKS

Classification of mollusks is a very confusing process, involving a bewildering array of colors, shapes, sizes, and types. In fact, there is no agreement within the scientific community as to exactly how many classes of mollusks there are. One of the most easily recognizable classes to novice collectors is the bivalves, which include creatures that produce shells of two equal parts. Oysters, clams, and mussels are among this group.

Bivalves consist of two shells that are hinged together. Each shell is a "valve," which translates to one panel of a folding door in Old English. *Kevin Adams*

Classification of mollusks is primarily based on differences in the foot and the shell, but these creatures are also identified by their teeth, DNA, diet, reproduction, and ecology. The foot, a single muscle that provides locomotion for mollusks, is described in detail in Chapter Three. This book focuses on the following five most-agreed-upon classes of mollusks.

Class Polyplacophora (Chitons)

Chitons are rock-climbing marine mollusks that live in abundance along rocky coasts throughout the world and down to depths below 1,500 feet. Among the most primitive mollusks, chitons are very simple organisms. They are bilaterally symmetrical with a long, flat body that begins with a mouth and ends with an anus.

A chiton's shell is responsible for its nickname: the armadillo of the sea. The shell consists of eight arched, overlapping valves or plates covered by spines, scales, beads, and a leathery sheath. Seen in a group on their rocky habitat, chitons look like miniature armored tanks with their overlapping plates protecting them from predators and the elements.

Typically, chitons clamp onto rocks or pilings and then hang on for dear life. They use their wide

Mollusks aren't restricted to water. Terrestrial snails are found in gardens and other upland habitats throughout the world's landmasses. *Shutterstock/Nicholas James Homrich*

Chitons are among the most primitive mollusks and live in rocky coastline areas. *Fotolia/Kerry Werry*

waste materials. The tube of a tusk shell is lined by a mantle that serves to absorb oxygen, since tusk shells do not have gills. The several hundred known species of tusk shells are distributed worldwide.

An adult tusk shell moves and searches for food with its buried head. The head has tiny sense organs called statocysts that detect food, such as forams (single-celled, shelled organisms), detritus (decaying organic matter), or an occasional buried bivalve. Once the food is located, *captacula*, thread-like cilia-bearing tentacles, bring the food to the mouth where a large radula grinds it up.

The sexes of tusk shells are separate and fertilization occurs externally at sea. The female releases her eggs one at a time. Each hatchling goes through a free-swimming larval stage before settling into an adult form on the muddy sea floor.

Class Bivalvia (Bivalves)

Bivalves include some of the world's best-loved culinary delights. Where would a coastal restaurant, bar, or tavern be without Oysters on the Half Shell, Clams Casino, Oysters Rockefeller, or Coquilles St. Jacque? Best of all, while these succulent delights tend to be quite pricey when served at local hang-outs, folks in the know can saunter over to nearby tidal flats and, within a few minutes, harvest a free bucket of steamer clams big enough to feed an entire neighborhood.

Including clams, cockles, mussels, oysters, and scallops, bivalves consist of two shells that are hinged together. Each shell is a "valve," which translates to one panel of a folding door in Old English. Each shell pair is linked by an elastic ligament and operated by incredibly strong adductor muscles. If you've ever tried to pry open an oyster or a clam, even with an official oyster knife, you know just how strong these muscles are. Bivalve shells have three distinct appearances: concentric sculpture (clam), radial sculpture (scallop), and latticed sculpture (oyster). Inside the protective hinged shells, the *visceral mass* (soft-bodied tissues) features a hatchet-shaped *foot* and a *mantle* sided by a pair of gills (see Chapter Three).

sole-shaped feet like big rubber sucker disks; once they are locked down, they are almost impossible to pry loose. When forced to move, chitons send waves of contractions through their oversized feet, pushing their bodies forward across their rocky substrate habitats. Yet there will never be any demand for chiton racing. Compared to a chiton, a sloth rockets through the treetops in which it lives. In a good year, a chiton may move a grand total of 10 feet.

To feed, chitons crawl along rocks and use their *radula* (rasping tongues; see Chapter Three) to scrape algae off their rocky habitats. Their teeth are sharp enough to etch glass.

Class Scaphopoda (Tusk Shells)

If ever a creature fit the description of being "just an old stick in the mud," it's the tusk shell. These innocuous mollusks have slightly curved cone shapes, which look like tiny elephant tusks. Their scientific name means "shovel foot" because they burrow headfirst into the mud on the ocean floor. They stay there with their posteriors sitting just far enough above the mud to siphon in water with oxygen and nutrients while flushing out water containing

The abundance of these large, glossy bivalves—sunray venus clams (*Macrocallista nimbosa*)—along the coast of northwestern Florida spawned a commercial fishing industry about 25 years ago.

The foot of bivalves varies widely, being well developed in the burrowing species (clams) and almost nonexistent in scallops, which "swim" to avoid trouble. The foot of a bivalve also has great relevance for seafood eaters. Steamer-lovers worldwide will recognize the foot as the built-in handle used for dunking clams in brine water and hot-drawn butter just before dropping them down the gullet.

Collectively, bivalves form a very important *trophic level* in the oceanic food pyramid. They provide a sustaining food source for fish, shorebirds, and gastropods, as well as augmenting the diet of many humans. Bivalve shells are also a major source of calcium for birds.

Bivalves are filter feeders, meaning that they nourish themselves by filtering water through their gills and gleaning organic matter that is suspended in the water. The filtering process is driven by a miniature pumping setup. Each bivalve has two siphons located in the rear portion of the shells. One siphon pumps water with oxygen and nutrients in, while the other takes wastewater out. Since they are filter feeders, bivalves do not have the rasping tongues, known as radula, found in most gastropods and other mollusks.

Most bivalves either attach to something or use their strong, sturdy feet to quickly burrow underground. Some mussels secrete byssal threads to attach to rocks. Oysters cement their left shells to a hard place, hence the need for seeding new oyster beds by depositing old shells at strategic locations on the ocean floor. This gives the oyster *spat* (free-swimming larvae) more places to settle down and develop into the adult bivalves that are so valuable to both marine and human food chains.

Class Gastropoda (Univalves)
The term gastropod is derived from two Latin words: *gastro*, which means stomach, and *pod*, which means foot. And stomach-foot is a fitting name for an animal that gets around by using a foot attached to the bottom of its abdomen. Totaling some 35,000 species, gastropods comprise more than one-third of all mollusks. While the best-known gastropods may be terrestrial snails and slugs, more than two-thirds of all species live in a marine environment.

Cultivation of the blue mussel *(Mytilus sp.)* occurs in more than 20 countries throughout the world. *Shutterstock/Hisom Silviu*

Characterized by having only one shell (hence their alternative name univalves), gastropods include snails, periwinkles, conchs, whelks, limpets, and sea slugs. Their single shells are typically spiraled, consisting of a coiled tube that increases in size as it winds around a central axis. Incredibly, certain gastropods instinctively build their shells to a logarithmic spiral known as the magic ratio. By doing this, their shells are regularly spiraled as they grow larger and larger around their central axes.

Gastropods have fleshy bodies with four clearly defined regions: the head, the foot, the mantle, and the visceral mass. The head includes sense organs (tentacles and eyes) and a mouth armed with a radula and sometimes jaws. The gastropod mantle often

Bivalves consist of two shells that are held together by a strong, internal adductor muscle. *Kevin Adams*

THE FIBONACCI SEQUENCE

In 1202, Leonardo Fibonacci discovered a mysterious series of numbers that is significant in art, architecture, oceanography, botany, biology, astronomy, and music. The Fibonacci Sequence is a mathematical sequence in which each number in the series is the sum of the two preceding numbers (e.g., 1, 2, 3, 5, 8, 13, 21).

If you divide a Fibonacci number by the next highest number, you will always get the ratio of 0.618034. The proportion of 0.618034 to 1 is known as the golden mean, and some gastropod mollusks instinctively build their shells to this logarithmic spiral known as the magic ratio. This proportion is also the mathematical basis for the shape of Greek vases, the Parthenon, playing cards, sunflowers, and the great galaxies of outer space.

includes the respiratory siphons. Due to an anomaly called *torsion* found only in gastropods, the digestive tract doubles back on itself so that the mouth and the anus are at the same end of each animal.

Mythological or not, Sasquatch, Big Foot of the Pacific Northwest, could never even compete with the conchs and whelks of the world. Gastropods rely on their oversized feet to move around, often with surprising quickness (see Chapter Three).

Some gastropods are also the 800-pound gorillas of the molluscan world. They viciously kill, drill, and suck the insides out of defenseless bivalves. The whole scenario can be compared to a classic David-and-Goliath story wherein the cute, cuddly scallop must jet around all over the place to escape the thunderous, incessant plodding of the gastropod.

Many world-famous gastropod shells are a collector's dream. Of particular note, volutes and cone shells are considered natural treasures of the deep. To most conchologists, these shells are far more valuable than any pirate's booty.

A favorite of children, the kitten's paw (*Plicatula gibbosa*) is a bivalve readily found in the intertidal zone on Florida's Gulf Coast.

Because of their popularity served on the half shell, oysters and clams are the best-known bivalves. *Shutterstock*

Worldwide, cone shells demonstrate breathtaking variability in colors and patterns. The glory-of-the-sea cone (*Conus gloriamus*) is one of the world's most celebrated seashells and even has its own Web site: www.gastropods.com/3/Shell_783.html. Found in the western Pacific, the glory-of-the-sea cone's value dropped significantly in 1969 when divers found more than 100 live specimens at Guadalcanal in the Solomon Islands. Even though the glory-of-the-sea cone is now affordable (about $25 on the low end) to serious collectors, its mystique and exquisite beauty still make it a treasure to behold. The precious wentle-trap *(Epitonium scalare)*—its name coming from the German word for spiral staircase, *wendeltrep*—is one of the most valued mollusks on earth because of the pattern of ribs on its back. Its turreted shell, consisting of whorls that form a high, conical spiral, has deeply ribbed sculpturing. Most wentletrap species are white, are less than 2 inches long, and exude a pink or purplish dye. Wentletraps occur in all seas, usually near sea anemones, from which they suck nourishment.

Class Cephalopoda

Cephalopod means "head-footed" and describes the group of mollusks that have their arms connected directly to their heads, bypassing their bodies altogether. Squid, cuttlefish, and nautiluses are all cephalopods and prove that cephalopods are among the world's most unusual-looking animals.

In general, a cephalopod looks like it was designed by committee and, as usual with committees, nobody could agree on anything. So the cephalopod parts appear to be randomly arranged by default. If you think this sounds pretty strange, you're right. Think about an octopus; it is one of the wackiest-looking creatures on earth with its head mounted in the middle of a whirling dervish of eight arms.

So, what about the shell in cephalopods? Aren't mollusks all supposed to have shells? Most cephalopods do indeed have shells. They just aren't readily apparent. In fact, you can't see them at all since they are internal, supporting cephalopods from the inside much like a human's calcium-laden skeletal bones support him from the inside. This evolutionary shift

Also known as the architecture shell in Germany, this beautiful Common American Sundial (*Architectonica nobilis*) is a gastropod that features spectacular spirals.

RIGHT
Found abundantly after storms along Florida's Gulf Coast, common figs (*Ficus communis*)—thin but stout gastropods—live in sandy areas and feed on echinoderms, especially sea urchins.

BELOW RIGHT
As their name implies, auger shells are all narrow and elongated with as many as 50 whorls. The Atlantic auger (*Terebra dislocata*), shown here, commonly lives near burrows of the acorn worm and probably preys on it.

away from species with large external shells, toward faster-swimming internal-shelled species, was most likely in response to competition from other marine animals for limited prey.

Contrary to popular opinion, an octopus' feet are called arms, *not* tentacles. Another very significant anatomical distinction among cephalopods, which differentiates them from their molluscan cousins, is that they have pretty standard body parts, albeit hooked up in unusual ways.

Cephalopods are the most intelligent invertebrates in the sea. With the largest and most complex brain of all mollusks, some octopuses and squid have actually been taught to do memory-dependent tasks.

Very good swimmers and active carnivores, cephalopods use their arms and tentacles to catch and hold quick-swimming prey. Cephalopods also have evolved some clever buoyancy mechanisms, expelling water through a funnel attached to their heads. This allows them to move around at will—up, down, and sideways—throughout the water.

Some cephalopods spend most of their time near the surface, while others are bottom dwelling. The giant squid is the premium example of a bottom-dwelling cephalopod, ruling over just about all that it sees as it roams the deep, dark depths of the ocean. There is, however, one gigantic exception to the giant squid's dominion over the depths (see Chapter Seven).

Conch shells like this scorpion conch (*Lambis*) are excellent shells for beginning conchologists to look for, since they're relatively easy to find and they're always spectacular. *Shutterstock/Alex James Bramwell*

Cephalopods exhibit a variety of defense mechanisms. Squid travel in schools and are very fast swimmers, allowing them to dart around like minnows and make quick turns and spurts to avoid predators. Octopods shoot out blobs of "ink" through their rectums, providing a smokescreen for escape. The brown or black cloud is inky because it contains a lot of a common animal pigment, melanin. In fact, the original sepia ink comes from the ink sacs of cuttlefish, which are close relatives of the octopus.

The ink may also contain some chemicals that poison the smell receptors and help discourage a predator. Some cephalopods have salivary glands that produce poisonous venom that can injure, and even kill, humans.

Cephalopods also have a phenomenal ability to change color, using pigment cells called chromatophores. When a cephalopod senses danger, it will change color—literally in the blink of an eye—to blend into the dominant background colors of its habitat. This process, known throughout the natural world as *cryptic coloration*, allows octopods and squid to virtually disappear from predators, avoid being eaten, and live another day.

Out of water, cephalopods are like limp dishrags, flaccid and gooey-soft to the touch. But in their oceanic element, they are beautiful swimmers, always appearing quite graceful and alert.

Cephalopods feed primarily on fish, shrimp, crabs, and other cephalopods. Their nervous systems are much more highly developed than those of other mollusks.

RIGHT

Although cephalopods—octopuses, squid, and cuttlefish—are true mollusks, their shells are mostly internal to their outer bodies. *Shutterstock/Tom Robbrecht*

A whirling dervish of eight arms connected to a central head, the octopus looks like a creature that was put together by committee. *Shutterstock/Rena Shild*

20,000 Leagues Under the Sea

Since it always seems to be playing on cable movie channels, I suspect most of you have seen the 1954 Oscar-nominated movie *20,000 Leagues Under the Sea*. Three things, all featuring cephalopods, about this masterful production will always stick in my memory. First is the name of author Jules Verne's famously futuristic watercraft, *The Nautilus*.

Then, who can forget the harrowing deep-sea battle between Captain's Nemo's *Nautilus* and the seemingly endless tentacled sea monster that kept capturing and dragging his sailors to a watery grave? As I recall, although the exact scientific name of this creature from the deep is not revealed during the movie, it is implied that the *Nautilus* is being attacked by a giant octopus. Since the largest known octopus is 8 feet long, including the length of its arms, it's more likely that the crew of the *Nautilus* is actually being attacked by a giant squid. Giant squid, reaching documented lengths of up to 70 feet, *do* live in the deepest parts of the world's oceans, another fact that dovetails with Verne's story.

Finally, I remember my dad laughing heartily when Kirk Douglas casually inquired about the meat course during an evening dinner with James Mason's Captain Nemo. Nemo answered, "Why it's boiled baby octopus, sir, how do you like it?" Douglas' face immediately curled into a gosh-awful expression looking like he had just been told that he had swallowed rat poison. Of course, back in the mid-1950s, eating a cephalopod sounded about as appealing to an American as eating a cockroach. Isn't it amazing what a few decades can do? Today, it's unusual *not* to see fried, broiled, and even raw octopus on a restaurant menu.

They depend on well-developed eyesight to locate their prey, which they then consume through a mouth and a parrotlike beak located at the base of their arms.

Cephalopods also are capable of sustaining high metabolic levels. In keeping with their strange anatomical arrangement, cephalopods have two kidneys and three hearts through which they pump blood that looks blue when it touches air.

When most people think about an octopus, they generally envision flopping, snakelike tentacles lined with rows of suction-cup suckers that latch onto prey. For the record, an octopus has 8 *arms*, none of which are tentacles. Meanwhile, a squid has 10 arms, 2 of which are extra long and are used for grabbing prey. These extra-long squid arms are the only true tentacles in the molluscan world.

Without question, the most curious and awesome cephalopod is the chambered nautilus (*Nautilus pompilius*). Unchanged since the Cambrian period—some 550 million years ago—the nautilus is the only cephalopod with an external shell. The distribution of the chambered nautilus covers the Andaman Sea east to Fiji and southern Japan and south to the Great Barrier Reef.

Visually, the nautilus is a stunning animal with a creamy white shell adorned with broad rust-colored stripes. The inside of a nautilus shell features an iridescent mother-of-pearl coating, while the outside is nature's most spectacularly proportioned sea spiral.

The animal's coiled shell is divided into chambers, each of which acts as a ballast tank to keep the body afloat. The brain controls the shell's buoyancy by actively pumping seawater into and out of the individual chambers.

A nautilus grows by adding chambers to its shell. In another amazing animal fact, the body of a nautilus always moves into the last and largest chamber added to the shell. After moving, the nautilus closes off the vacated home chamber with a layer of mother-of-pearl.

Of all the mollusks, the squid is the only one that can truly grow to monstrous proportions. Giant squid up to 59 feet long have been documented, with speculation that there may be squid that exceed 100 feet in the deeper parts of the ocean. *Shutterstock/ Aleksandrs Marinicevs*

The chambered nautilus *(Nautilus pompilius)* is celebrated worldwide for its exquisite beauty and its unique physiology. *Shutterstock/Michael Fuery*

Chapter 3

Spineless Wonders
THE ANATOMY OF MOLLUSKS

Remember seeing your uncle eating steamed clams, "steam-ahs," during that family trip to Maine? First, he dug out the fleshy, amorphous blob of meat. Then he held the mass, dripping with brine and butter, over his open mouth and gleefully dropped it in. He then gulped down a slug of cold beer. Well, if your uncle took the time to look closely at the clam's body, he might not be so excited about eating it. In fact, he might decide not to swallow at all.

On close examination, the fleshy mass of a steamed clam has all the basic animal body parts (heart, eyes, reproductive organs, etc.), plus a few extra things. Most of you would probably never dream of gulping down an entire fish, yet that's precisely what you do when you fork, dip, and swallow oysters and clams. If you're a raw bar fan, I apologize for spoiling your appetite. I just thought you should know.

PRIMARY MOLLUSCAN BODY PARTS
While axioms like "true beauty is found only on the inside" and "you can't judge a book by its cover" are often cited when dealing with human appearance, nothing could be further from the truth in the world of mollusks. The word *mollusca* is Latin for soft and that's exactly what you find when you open a shell.

The Visceral Mass
Sounding like a name concocted by a Hollywood horror flick producer, a mollusk's visceral mass is entirely nonmuscular. It contains all of the bodily organs for digestion, circulation, reproduction, and respiration. Unique to all mollusks, the mantle is also part of the visceral mass.

The Mantle
The mantle is found only in mollusks, nowhere else in the animal kingdom. In bivalves, the mantle consists of two flaps of body tissue that are attached to the dorsal (top) surface of the visceral mass. Also called the skirt or *pallium*, the mantle typically hangs freely down both sides of the visceral mas.

Depending on the species of mollusk, the mantle is a lobe, a pair of lobes, or a fold of muscular flesh containing specialized glands. These glands convert salts in the mollusk's blood to a liquid form of

The word *mollusca* is Latin for soft and that's exactly what you find when you open a shell. *Shutterstock/Tina Rencelj*

The most common non-marine mollusks are freshwater and terrestrial snails. They're found everywhere, from the tiniest creeks to backyard gardens. *Shutterstock/Vaide Seskauskiene*

calcium carbonate. The cells at the edge of the mantle then secrete a calcium-laden liquid. Next, the liquid solidifies, forming a shell.

As the visceral mass of a mollusk grows, additional shell production is required for support. This is accomplished when the mantle spreads another layer of liquid calcium carbonate onto the lip of the shell. Since the thickness of each layer of calcium carbonate is slightly different, growth lines form as a shell gets larger. These lines are very prominent on certain shells like the Atlantic surf clam (*Spisula solidissima*) and the knobbed whelk (*Busycon carica*).

Each shell that you find on the beach shows a pock mark or indentation indicating where the mantle was attached. Just like human fingerprints, each shell is unique to each individual mollusk species. Accordingly, examining shell type and structure are the primary ways to identify mollusks.

Each mantle encloses a mantle cavity containing fluid that is continually replaced with outside water. This fluid turnover carries excess water, ions, and wastes away while helping circulate nutrients and oxygen throughout the visceral mass. The mantle is also essential for respiration since a mollusk's gills or lungs are housed in the mantle cavity.

The Foot

Yes, it's true that 70 percent of mollusks have only a single foot. Ah, but what a foot it is. Shaquille O'Neal's size 18 sneakers? They are just small potatoes in the molluscan world. A mollusk's foot dominates its body, typically being more than 50 percent of its body length. In human terms, that would mean a normal-sized man would have a 34-inch foot. Sorry, Shaq, but you're not even close on this one.

Being the most muscular part of the body, the molluscan foot is adapted for locomotion in a variety of ways (see Chapter Five).

The Digestive Tract

Hard as it may be to believe when looking at an oyster on the half shell, all mollusks have a complete digestive tract, starting with the mouth and ending with an anus. The anus empties into the mantle cavity, which also receives waste pumped in from the kidneys.

The Siphon Tubes (Bivalves Only)

Clams and some other bivalves separate their shells and extend two siphon tubes above the sand. One tube draws in water containing oxygen, minerals, and plankton, while the second tube pumps out waste material.

Hinge and Teeth (Bivalves Only)

Bivalves have a type of trap-door arrangement that allows them to snap shut whenever danger lurks nearby. Look closely at the inside of a clamshell and you'll see a prominent scar. This is where the snapping, or adductor, muscle was attached.

   The *teeth* of a bivalve have nothing to do with eating. They grow on the interiors of each half shell and interlock when the muscles snap the hinge shut. If television producers ever come up with a show entitled *CSI: Outer Continental Shelf,* investigators would spend a lot of time studying bivalvean teeth, since the teeth are unique to each individual species of bivalve.

Radula (Gastropods Only)

A mollusk's *radula* is found nowhere else in the animal kingdom. It is specially adapted to various feeding techniques in different gastropod species. A radula is mostly composed of chitin, the tough, horny substance that forms the outer covering of insects and crustaceans, such as crabs and lobsters. Visualize the radula as a human tongue, but instead of soft, it's very tough. Its sandpapery surface is used for ripping and shredding foodstuffs right off the ocean floor. Like a built-in food processor, a radula cuts, slices, dices, and chops to suit the preferences of the gastropod it serves.

In the living bivalve, muscles from each half shell attach to the *hinge,* which holds the two halves together. *Shutterstock/Erica Chard*

Most ceriths (family Cerithiidae) live in rubble mixed with sand. These gastropods all feed by using their radulas to scrape algae and detritus off neighboring rocks.

Cephalopods have very different features than most other mollusks. In addition to having an internal shell, they also have a highly developed nervous system and functioning eyes.

Circulatory-Respiratory System

All mollusks, except cephalopods, have open circulatory-respiratory systems, complete with a heart, blood vessels, and respiratory pigments. The basic molluscan respiratory system works like this: Beating cilia (tiny hairs) in the mantle cavity generate water currents that draw water in along both sides of the mantle. Once inside the mantle, the water passes over the gills where oxygen is taken in and carbon dioxide is released into an outward-bound current that flows between the two inward-bound currents.

Nervous System

If someone told you that the oyster on the half shell you're about to consume is a very sensitive animal, you would probably just laugh in delight while continuing your meal. The truth is that mollusks do have relatively complicated nervous systems with a pair or

The banded tulip (Fasciolaria lilium) is one of the most aggressive gastropods, often consuming its close cousin, the true tulip (Fasciolaria tulipa), with its powerful radula. *Kevin Adams*

This lightning whelk's (*Busycon contrarium*) unusual leftward-turning spiral gives it an advantage against predators.

The pear whelk (*Busycon spiratum*) is distinct from its lightning whelk cousin because it is "right-handed"— or normal—and less knobby around the "shoulders."

more of ganglia just behind the mouth. The nerve ring surrounding the ganglia gives rise to two primary nerve cords that spread throughout the body, transmitting sensitivity to touch to each part of a mollusk.

Vision

Although vision is poor in most mollusks, cephalopods—such as octopuses and squid—have eyes with lenses, retinas, and other features remarkably similar to human and other vertebrates' eyes. The hawk-wing conch (*Strombus raninus*) has extremely well developed camera-type (simple) eyes that provide it with excellent night vision.

Operculum (Gastropods only)

When a gastropod recedes into its shell, it drags a piece of shell into place over the shell's opening, effectively slamming the door shut. This trap-door-like structure is known as the *operculum*.

Sinistralism

Left-handedness in nature is called *sinistralism*, meaning sinister—which, when applied to humans, is about as low as you can go on the desirability scale. This bias is seriously magnified in the molluscan world where left-handed (leftward-spiraling) shells are somewhat unusual. But to collectors, these leftward spiraling shells are rare, lucky finds. Among the shells commonly found in Florida, only lightning whelks typically have a left-leaning pattern.

These left-spiraling shells actually have an advantage when it comes to fending off prey. When Yale University geologist Greg Dietl put live snails into a tank with snail-eating crabs, the crabs used their sharp right claws like can openers on their prey. When the shell-crushing crabs picked up "lefty" snails, they often failed to open them and, after many tries, finally just gave up.

Chapter 4

Beauty Is Only Shell Deep
THE SKELETONS OF MOLLUSKS

Humans keep their bony skeletons on the inside while showing their softer sides to the outside world, but mollusks are the opposite. Their shells are the bony skeletons that they show the world. So how do these skeletons grow around the outside of these organisms? Modern medicine might be mind-boggling, but it's no match for nature's magic elixirs. Show me a doctor who can turn a mish-mash of flesh and organs, dominated by a single foot, into a perfectly sculpted seashell and I'll show you a true genius.

PRODUCTION OF SHELLS

Each mollusk generates its own shell as it grows. Shells are made of liquid calcium carbonate crystals and produced by the mantle. Calcium carbonate, a salt found in the blood of mollusks, is obtained either from the food mollusks eat or waters in which they live. Shell crystals are deposited in layers of varying size, shape, and orientation. The layered construction strengthens the entire shell while the formation of spines, grooves, and ribs on shells also adds strength in specific areas.

Bivalve shells have three layers for increased strength. The oldest, raised part of a shell is called the *umbo*. *Varices* are ridges showing where a mollusk paused to rest while constructing additions to its shell house. Interestingly, varices are one of the keys to determining a shell's value to collectors.

Timing for producing new parts of a shell depends on several factors, including:

• Sexual hormones
• Diet
• Acidity of water
• Water temperature

TYPES OF SHELLS

Research shows that a mollusk shell is remarkably durable; it can be stiffer than aluminum and stronger than fiberglass. The shell of the abalone is known for being exceptionally strong. It is made of microscopic calcium carbonate tiles stacked like bricks. When the abalone shell is struck, the tiles slide instead of shattering and the protein stretches to absorb the energy of the blow. Material scientists at the University of California, San Diego, are studying the tiled structure for insight into making stronger

Many gastropods instinctively create their shells in a logarithmic spiral with a proportion of 0.618034 to 1, known as the magic ratio. *Shutterstock/Alex James Bramwell*

Bright colors, complex patterns, and intricate appendages—as shown here on this *Chicoreus* sp. shell—make shell collecting a joy for both young and old.
Shutterstock/Andrew Williams

ceramic products such as body armor. If they are successful, they will be able to create concrete that can withstand 1,000 times the impact of ordinary concrete.

Shells are specially adapted to their environments. Each shell is designed to make life easier for its molluscan maker. That's part of the reason a mollusk shell can be a multiplicity of shapes, colors (rainbow hues), and patterns. A cowry shell is smoothly polished and highly glossy with bright intricate patterns. While exquisitely beautiful, these characteristics actually help cowries survive. Cowries' slick shell surfaces are impossible for crustaceans—such as crabs and lobsters—to grab with their pincers.

Other mollusk shells have lacy frills or elongated spikes. The spikes provide an extra layer of protection, much like the quills that armor slow-moving porcupines. Tusk shells are streamlined for burrowing into soft sand, while wentletraps have heavy ridges that serve as anchors in bottom mud. Still other mollusks like the king and queen crowns grow spines that entrap and encourage the growth of camouflaging seaweed and corals.

Shells often tell the life stories of their "owners." Breaks and chips often mean a mollusk has survived many battles with predators. Color changes indicate changes in water chemistry. Thickened edges and dull colors are signs of old age in mollusks, like osteoporosis is in humans.

COLORS AND PATTERNS OF SHELLS

Even world-famous art museums, like Paris' Louvre, London's Tate, or New York City's Whitney, find it difficult to compete with the colorful patterns found in the natural world. In fact, throughout history, great artists have been awed and inspired by what they have seen during walks along the beach.

Molluscan shell colors are innate in each species with some natural variation due to environment and diet. For example, frilled dogwinkles are normally gray or white, but they turn yellow after they are fed diets of barnacles and purplish-brown when they eat blue mussels.

Organic pigments in nutrients (food) are processed by the mollusk, then distributed by the blood system and mixed with liquid calcium carbonate before the shell hardens. Four main pigments produce

Seashells are made of liquid calcium carbonate crystals and produced by the mollusk's mantle. *Shutterstock/Maria Menshova*

Mollusk shells exhibit every surface imaginable, from smooth and glossy to deeply furrowed—shown here. The variations in shell type match each species' special habitat conditions. *Shutterstock*

The auger and turret (Turritellidae) families have narrow and elongated shells with up to 50 whorls. Many of these shells are also host-specific, meaning that they feed entirely on one kind of marine worm. *Shutterstock/Alex James Bramwell*

The slick shell surfaces of cowry shells are almost impossible for predators to grab. *Shutterstock/ Nikolay Okhitin*

colors in shells: *carotenoids* produce yellow; *melanins* produce black; *porphyrins* produce green; and *indigolds* produce blue and red. Shells buried in sand or mud absorb the color of their substrate, while long exposure to the sun bleaches shells white. Most color cells are located along the front edge of the mantle, and this is where new shell material is added. When color cells remain in the same position as the shell grows out, straight color lines result. Patterns of dots and dashes are produced when pigment production continually starts and stops. If the color cells migrate to one side or the other, slanting trails of color are produced. Migrating color cells yield all sorts of artistic variety, including forming circles, squares, and triangles.

SHAPES OF SHELLS

In the human world, you can tell a lot about the local climate by looking at how houses are constructed. In the Snowbelt, A-frames and other home styles with steeply pitched roofs are the trend. This design uses gravity to keep rooftop snowpacks to a minimum and eliminates the daunting task of shoveling several feet of snow while perched 20 feet above the ground during a howling blizzard. Meanwhile, Sunbelt homes have thickly insulated, flat roofs to minimize the space that has to be cooled with year-round air conditioning.

Similarly, shell shape can tell a great deal about the living environment of a mollusk. Low, wide shells often reflect an area of strong currents and many predators. Emphasizing shell width over height gives mollusks in these habitats the ability to hold tight to their substrates while minimizing visibility and the need to flee from predators. Flat, saucerlike shells may indicate areas where the sea floor is hard, emphasizing the need for retreating into the shells, instead of burrowing into soft sand or mud. Thin,

The long, thin shape of the razor or jackknife clam (family Solenidae) allows it to burrow deep into mud flats where it often lives in permanent burrows. *Shutterstock/Miranda Zeegers*

If there was a beauty contest for mollusk shells, the chances are good that the winner would be from the volute, wentletrap, or cone—shown here—families. *Shutterstock/Jubal HarshawSamarin*

The murex family of mollusks is best known for long, lacy spines, vividly displayed by this venus comb (*Murex pecten*). *Shutterstock/Alex James Bramwell*

spherical shells may show the need for deep, cold-water homes, like those around the North and South poles. Smooth, long, and tapering shells are typically designed for burrowers that live in habitats with soft, sandy-bottom surfaces.

In nature, size really does matter. A threatened animal always wants to make itself look as big as possible. To accomplish this, birds fluff out their feathers, cats arch their backs, and crabs stand as tall as possible while brandishing their fighting claws in front of their bodies.

Mollusks are no exception to the "bigger is better" rule. Those living in areas of calm water grow spines and other sculpted appendages that increase the volumes of their shells and make them appear more formidable. The increased shell volume also provides additional surfaces for the attachment of other marine organisms, allowing a mollusk to better blend into its sea floor environment. This camouflaging process is widely used throughout the animal kingdom, especially for protecting eggs and young. Think about how well the mottled brown, black, and white feathers of a piping plover chick match the colors of a sand dune ridge. And how often have people been startled after nearly stepping on a white-spotted fawn lying motionless in the sun-dappled light of an oak-hickory forest?

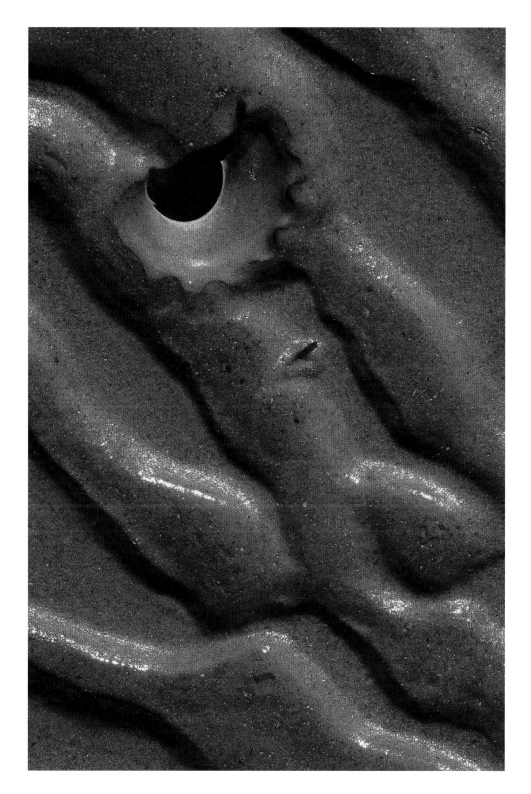

Moon shell (family Naticidae) females embed egg capsules into thin, flattened spiral masses of sand grains. These are the "sand collars" frequently seen in the intertidal zones of beaches. *Kevin Adams*

Chapter 5

The Good Life Starts at the Bottom
MOLLUSCAN LIFE HISTORY AND BEHAVIOR

To really understand mollusks, you need to don diving gear and explore the bottom of a coastal shelf. Here, you'll find a vast, living community of mollusks, proudly displaying a sunken garden of shells of every imaginable shape, color, and pattern. You could also learn a little bit about the behavior of mollusks if you were able to stay underwater for a few weeks and see how they really live.

REPRODUCTION

Fecundity describes an animal's ability to successfully reproduce or replace itself. When an animal species cannot successfully replace itself, the abundance of that species—its total population on earth—will begin to decline. If the decline is too rapid, the animal species will soon reach the point of no return, past which there is only extinction—the loss of that species forever.

In the 1800s in the United States, the passenger pigeon and the Carolina parakeet both occurred in such abundance that their migratory flights regularly darkened the skies over the Midwest and Southeast. But due to habitat loss and overhunting, their populations declined so rapidly that they were both declared extinct by the early 1900s. As the result of passing the U.S. Endangered Species Act (ESA) of 1973, other American species were saved from such a fate, including the bald eagle, the peregrine falcon, and the timber wolf. Operating with the regulatory hammer provided by the ESA, wildlife biologists were able to save these species by protecting their critical habitats while increasing their fecundity, or successful reproduction. The peregrine falcon was saved by hacking, or removing eggs from captive birds to cliff-top nest sites of wild falcons.

Generally speaking, the larger the animal, the lower the number of offspring is needed to ensure survival. This means that a mating pair of grizzly bears has a low fecundity rate, needing only to produce two cubs to successfully replace themselves, since the odds of a grizzly bear cub reaching adulthood are very high. On the other hand, the odds of a green frog tadpole reaching adulthood are extremely low. Tadpoles are very small and defenseless little packets of protein just waiting to be eaten. So, in order to simply replace themselves on earth, a mating pair of green frogs has to produce hundreds of fertilized eggs. Thus, the pair of green frogs has a very high fecundity rate.

True conchs (*Strombus* spp.) are herbivores—algae eaters—and are often seen grazing in large groups like cattle in a field.

With the exception of cephalopods, mollusks generally populate the small end of the size spectrum. So, following the rule of fecundity, each individual mollusk must produce lots of young to successfully replace itself on the planet.

The lightning whelk, a gastropod, produces strange-looking strings of quarter-sized, nut-brown discs that often show up on beaches. Looking like pieces of some sort of primitive necklace, these discs are the whelk's egg cases. If you break one open, you'll find that each disc contains as many as 100 tiny whelk shells.

Lightning whelks have separate sexes, and they join to mate. A female lightning whelk fertilizes her eggs internally and then begins to spawn her egg cases six weeks after mating. It then takes her 8 to 12 days to spawn 24 inches of egg cases. On average, a 9-inch whelk will produce a 30-inch-long string of cases. Each disc of the string contains 30 to 100 baby whelks in tiny shells.

Large strings may contain a total of 5,000 to 12,000 young, of which only a few will survive. Although this number seems high, it's actually only a moderate fecundity rate. This moderate fecundity rate makes sense, given that whelks are medium-sized mollusks.

The majority of mollusk species are *dioecious*, meaning that they have separate male and female sexes (although some species are *hermaphrodites*, having both sexes in the same animal). Still other species change sex one or more times during a lifetime.

Tusk shells, chitons, some gastropods, and most bivalves discharge large numbers of eggs and sperm into the water. Random fertilization occurs, either externally in the sea or internally inside the mantle cavity. Fertilized eggs then produce free-swimming larval stages, called *trochophore* and *veliger* larvae. It's in these larval stages that mollusk babies are most vulnerable. They become part of the collective mass of organisms known as *plankton*, the surface-floating buffet table that forms the wide base of the food pyramid in the world's oceans. Starting life as floating hors d'oeuvres does not engender a high survival rate. Since mollusk parents do not nurture or protect their young, successful reproduction depends on the sheer volume of larvae produced. One 4x5-inch oyster can produce 50,000,000 eggs in a single year. Now that's a *high* fecundity rate—producing fifty million young just to replace yourself!

Colorful seashells surround two whelk egg cases on a beach on Sanibel Island, Florida.

Even with fecundity rates in the millions, only a fraction of a percent of mollusk babies successfully reach adulthood. When a tiny mollusk hatches from its egg, the larvae come into the world equipped with tiny shells. As the larvae grow, the shells—via calcareous secretions from the mantle (see Chapter Four)—grow with them. The extremely lucky ones, literally those one in a million, survive larval life and settle to the bottom to live as adults.

Cephalopods do not go through the traditional larval stage of development. Cephalopod reproduction occurs when the male produces a special arm called a *hectocotylus*. With this, he inserts a sperm packet into the larger female's mantle cavity. The female later lays a number of large, yolk-filled eggs, from which the babies hatch. More developed cephalopods, such as octopods, guard their eggs. They keep them clean, provide them with fresh water, and defend them against predation. The female octopus does not eat while she is guarding her eggs. She often dies after the eggs hatch, having become too weak to defend herself from predators.

GETTING AROUND

As discussed in Chapter Three, the molluscan foot is huge, especially when compared to the feet of most other animals, including humans. Mollusks need oversized feet because they use them as both a backhoe and anchor. When a mollusk senses danger, it uses its foot to rapidly dig into sand or mud to escape

Imagine this Atlantic Triton's Trumpet (*Charonia variegata*) dragging itself along the ocean floor, and you'll quickly understand what "working at a snail's pace" really means. *NHPA/Pete Atkinson*

While its name may connotate peace and joy, the true tulip (*Fasciolaria tulipa*) is a rugged carnivore, greedily chowing down on bivalve mollusks.

Many bivalves take in food and eliminate waste by extending siphon tubes like this one up above the surface of the ocean floor. *AGPix/Jeffrey L. Rotman Photography*

potential predators. When strong currents roil and rip the ocean bottom, a mollusk sets its foot down deep and tight, much like a ship using its anchor to brace for an oncoming storm.

Whelks, conchs, and other gastropods move by contracting the foot muscle in a regular, wavelike ripple that inches them along with extreme slowness. Given this mode of moving about, it's easy to understand the derivation of the phrase "working at a snail's pace."

When it comes to locomotion, scallops are the Keystone Kops of the ocean bottom. Lacking feet, they use the exact opposite approach of their molluscan brethren. They rely on a type of jet propulsion to stay out of harm's way. By clapping their bivalve shells together, they lurch quickly and wildly all over the place. The result is a helter-skelter effect that is startling and funny. Watching them in a laboratory tank, you can quickly understand how the sudden sight of scallops spurting off in all directions would both confuse and frustrate a would-be predator.

BREATHING AND FEEDING

Clams have an infamous reputation for "spitting" on people walking through their beds. In reality, they are just trying to catch their breath, simultaneously drawing in clean water and discharging wastewater through their siphon tubes. If you're a clammer, you know that this breathing routine typically leads to a clam's demise, since each spout tells you precisely where to dig. Twelve spouts and you've acquired a sumptuous feast, ready to go on the half shell—with cocktail sauce, please.

Oysters, clams, and other bivalves are primarily sedentary filter feeders. As with the vast majority of animals in the ocean, bivalves primarily eat the plankton soup that saturates ocean water. Strangely, since mollusk larvae start out as free-swimming members of the plankton community, adult bivalve meals include consuming many of their own kind.

To both breathe and eat, a bivalve separates its two shells and extends its two siphon tubes up above the sand and into the water. One siphon then draws in water containing oxygen, minerals, and plankton,

Clams have a reputation for spitting at people who are walking through mud flats at low tide. In truth, the clams are breathing by taking in and discharging oxygenated water through their siphon tubes.

Found in abundance just below the sand's surface, the lettered olive (*Oliva sayana*) feeds on coquinas by wrapping the tiny prey in the hind part of its foot, then dragging them down under the sand for digestion.

while the second siphon pumps out waste material.

Completely different from their bivalve brethren, some gastropods are aggressive carnivores, meaning they prefer to chase down and dine on meat. In fact, they feed primarily on their staid, defenseless bivalve cousins. In a typical feeding attack, a whelk either forces a clam's shells apart with its foot or uses specially designed body parts to drill through the clam's shell.

Once attached, the whelk extends its *proboscis* (long snout) into the drilled hole and eats its clam cousin alive. Talk about families not getting along at the dinner table!

On the other hand, the fighting conch is surprisingly a vegetarian, using its radula to scrape algae off the ocean floor. Conchs often graze in large colonies, looking like herds of cattle in a sunken pasture.

Many mollusks exhibit some rather unorthodox table manners. For instance, the tiny nocturnal cone snail emulates the derring-do of nineteenth-century whalers. This trickster lies in waiting until its prey approaches. When the moment is right, the cone snail shoots out a tiny glass harpoon-like tooth. If the cone's aim is true, the tooth impales, poisons, and paralyzes the startled victim. Then the snail cone saunters over to its victim, usually a small fish, and swallows it whole.

The vampire snail (*Colubraria* sp.) of the central California coast also uses skullduggery to secure its nightly repasts. Lying camouflaged in the sand, vampire snails wait for a torpedo ray or angel shark to settle in nearby for the night. Once the night's meal ticket appears sufficiently immobile, the vampire snails emerge from their hideouts, bite into the soft gill and mouth tissues of their sleeping quarry, and drink their fill of blood.

LIFESPAN

As adults, most mollusks live for several years. While scallops typically live for two years, horse conchs (*Pleuroploca gigantea*) can live for as long as 15 years (and grow up to 24 inches). Lightning whelks typically grow to 16 inches long and live 10 to 12 years, with some whelks surviving for more than 20 years in captivity.

Chapter 6

So Where Were You Seeded?
Molluscan Habitats and Distribution

Buying a home can be an excruciatingly painful process. Everyone has an ideal place in mind, with specific features that mesh perfectly with his or her needs. Unfortunately, most of us end up settling for something that's less than ideal, as it's virtually impossible to find everything you're looking for at an affordable price.

Most animals go through a search for living space that is surprisingly similar to the human house-hunting experience. Every animal has a *preferred habitat*, a space with specific features that allow it to successfully live and reproduce. Consider the living space demands of a pack of timber wolves. They need a large tract of undeveloped forestland, called a *territory* or *home range,* to successfully live and breed. If the forestland also has a healthy population of *ungulates* (deer, elk, and moose), the wolf population will thrive, breaking off into new packs every year.

A NICHE IN TIME

In scientific terminology, an animal's preferred living space is called its *ecological niche.* The ecological niches of mollusks are extremely varied and widespread.

Mollusks can be found worldwide, from the deepest ocean trenches to the highest mountaintops and everywhere in between. While this book focuses on saltwater species, mollusks also inhabit every type of freshwater habitat in the world. Freshwater mussels live in most of the world's large rivers and major tributaries, while terrestrial snails are found everywhere from domestic flower gardens in the Midwestern United States to mangrove swamps along the U.S. coastlines. Look closely at a cluster of mangroves and you'll find snails everywhere, nestled among the exposed tree roots and clinging to the branches, in muddy, tight thickets.

In general, the best habitats for mollusks occur in tropical waters, and this is where the greatest variety and most spectacular shells are found. Tropical coral reefs are home to multihued rainbows of cones, cowries, and volutes, while subtropical swamps are literally crawling with oysters, snails, and small, globe-shaped univalves called nerites.

The world's temperate waters also provide preferred habitats for a wide range of mollusks. The great estuarine river systems of the eastern Atlantic

Since the advent of writing, many people have believed that the markings on alphabet cone (*Conus spurius*) shells are letters spelling out secret messages.

Although they vary in size, color, and shape, the crown conch (*Melongena corona*) colonies found in mangrove swamps from Florida to Texas are all members of the same species.

coast provide a mosaic of habitats, including incredibly rich tidal marshes that are referred to as the nurseries of our oceans because of the sheer volume of marine species that is produced in them every year. To the south, these massive estuarine systems are interlaced with broad sand beaches and coastal flats, while the northeastern shores of the United States feature an abundance of rock-bound coastlines honeycombed with species-rich tidepools.

In oceanic habitats, mollusks have adapted to survive in a variety of ways. Some species thrive inside mats of seaweed, while others live on the hold-fasts of coral or buried in the roots of undersea grass beds. Some species spend their entire lives floating on a raft of bubbles, while others are adapted to living near thermal features like hot springs. Some of the more unusual molluscan niches include the undersides of sea urchins, inside the gills of fish, and among the arms of starfish. Still other species burrow into living coral and grow as it grows, or they attach themselves to the rootlike bases of living coral reefs.

Among mollusks, limpets are the uncontested kings of severe habitats. They use their dome-shaped shells to withstand constant buffeting by wind and waves, securely clinging to their rocky habitats during even the strongest of storms.

MOLLUSCAN GARDENS

Many people believe that most mollusks live on beaches and along shorelines. Since this is where you find seashells, mollusks must live nearby, right? There is some truth to this assumption. First, many species do live between tide lines. Second, it's just a matter of timing, going back about 10,000 years to the last Ice Age. Today's classic molluscan neighborhood is a coastal shelf covered by 30 to 100 feet of water. During the last Ice Age, when much of the present ocean water was locked up in massive polar ice caps, the areas that are now coastal shelves full of mollusks were beaches and intertidal zones.

One of the best examples of a contemporary molluscan community is Florida's Outer Continental Shelf (OCS), a vast sweeping area of shallow water overlying oil-rich sediments that extends from the Alabama state line to the Florida Keys. More than 1,000 miles long and averaging 80 miles in width, the Florida OCS provides habitat niches for millions of mollusks. It's here—and on other coastal shelves and in tropical coral reefs throughout the world—that mollusks put on their greatest show. The sheer abundance and diversity of living mollusks in these undersea gardens is truly mind-boggling.

So if these off-shore coastal shelves are the prime hangouts for living mollusks, how and why do so many of their remains—their shells—end up covering beaches and coastlines? The answer is one word: storms.

Since Florida's Sanibel and Captiva Islands are east-west-oriented barrier islands, they provide the perfect situation for illustrating the movement of living mollusks during a typical storm event. When a winter storm hits the Gulf of Mexico, it brings strong northerly winds that push the surface water to the south. The southward-blowing surface water creates a volumetric void that in turn creates north-flowing bottom currents. These strong bottom currents swirl along, uprooting thousands of mollusks and dragging them up onto Sanibel and Captiva beaches where they become stranded and quickly die. Despite this, only a very small fraction of mollusks ever reach the beaches and shorelines where people are used to seeing them.

Keyhole limpets (*Diodora* spp.) are distinguished by radial ribbing that descends from dumbbell-shaped openings in their centers, making them look like the coolie hats worn by workers in Asian rice paddies.

Among mollusks, limpets are the uncontested kings of severe habitats. *Shutterstock/Geoffrey Whiting*

PURSUING THE ELUSIVE ANGEL WING

One of the best examples of knowing how to find and collect a specific type of shell is provided by my dad, Franklin H. Titlow Jr. A career newspaperman, he and his lovely wife, Elizabeth (Liz), spent many winter months collecting shells along Florida's Gulf Coast. Liz's favorite shell was the angel wing, an extremely fragile bivalve that perfectly emulates the ethereally white shape and pattern of its namesake. Here's my dad's description of collecting this delicate beauty:

Luckily for us, these exquisitely fragile bivalves occurred in relative abundance in the limited area where we puttered about This was in the Imperial River, just a strand of sand away from the shoreline of the Gulf. Getting an angel wing required our paddling along among the small white patches of sand showing just a few inches above the water. At certain places, I would look for a hole in the surface that indicated the burrow of an angel wing down below. To investigate further, I would get out of the boat, leaving Liz to hold her steady as I rolled over onto my side in the wet sand. Then I would start digging down, keeping some distance away from the breathing hole so as not to disturb and damage our quarry.

As I dug down very carefully, I would be up to my elbow before I felt like I was getting close. I was feeling for something that felt like a shell. But I had to be so very, very careful since I was down deep enough that it was getting fairly watery. And I sure didn't want to feel something kind of gooey, like the animal itself. No sir, I had to dig under him or I would risk breaking apart the fragile hinge that connected the two wings of

my quarry. The only way to successfully collect the intact wings was to come up from underneath and cradle the entire creature in the palm of my hand. As soon as my hand was in the proper cradling position, the real challenge began. I would gently start pushing the wet sand away from my hand, inching up onto my elbow, then to my knees, gently raising the gleaming angel wing's body to the sand's surface.

As a final step, I would ease the wet mass onto the dry sand so Liz and I could determine if I had managed to extract a complete angel wing. For the few times, among many tries, that we were successful, we ended up with several intact angel wings with no nicks or breaks in the fragile shells and, most important of all, the hinge still in place and holding the two wings together.

Looking positively angelic, these fragile bivalves (*Cyrtopleura costata*) must be handled with great care. Telltale holes in the mud reveal an angel wing's presence in tidal flats.

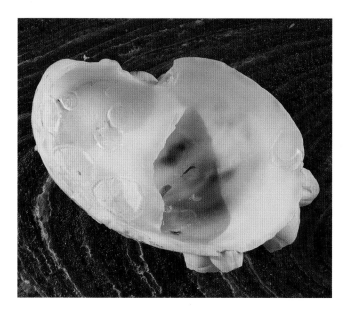

Often found stacked on top of each other in oyster beds, Atlantic slipper shells (*Crepidula fornicata*) have an interior half shelf that provides their namesake.

Actually a clam, the Atlantic bittersweet (*Glycymeris undata*) is thick-shelled and almost round. It typically lives buried in sandy bottoms with water depths ranging from 5 to 150 feet.

SHELLS WE SHOULD ALL KNOW

One of the better-known inhabitants of Florida's Gulf Coast is the fighting conch (*Strombus alatus*), which lives in the tidal mud flats of brackish-water estuaries. This impressive gastropod has points on its shell that are reminiscent of the spikes worn by Roman Gladiators. Belying its name and appearance, the fighting conch is primarily a docile, seaweed-eating animal. The fighting moniker derives from its habit of waving its operculum around when it's either picked up or flipped over.

The tiny Atlantic coquina shell (*Donax variabilis*) can be found in a whole rainbow of colors just below the sand's surface at the surf line of the open ocean. Children love coquinas because they are easily collected by digging into wet sand as a wave retreats down the beach. Boiling live coquina shells provides the stock for a popular chowder in Maryland and Virginia.

The uniquely patterned lettered olive (*Oliva sayana*) also resides in the sand at the surf line. The olive's name comes from the dark, tentlike markings on the shell.

Well-marked shells display the "lettering" in two broad bands, which are separated by a pale, unmarked band.

The turkey wing (*Arca zebra*) lives attached to rocks just off shore and is commonly found on beaches after being dislodged by storms. At first glance, collectors are amazed by this shell's resemblance to an actual turkey wing, down to its exact shape and brown-banded coloration.

One of the world's most strikingly beautiful shells, the king's crown (*Melongena corona*), also called the crown conch, is usually found on mud flats near mangrove swamps and oyster beds. A ring of spines surrounding its spiraling top gives this shell its name. A thorough cleaning reveals exquisite patterns, featuring prominent black and white bands garnished with traces of brown.

Florida's state shell, the horse conch (*Pleuroploca gigantea*) lives throughout state waters at depths ranging from 1 to 80 feet. Horse conchs, also Florida's largest shells, can grow up to 24 inches long. They feed very aggressively—usually on clams—and can even be cannibalistic when bivalves are in short supply.

Good finds for beginners, murex shells (family Muricidae) are relatively common and very showy.

The pink conch (*Strombus gigas*) is found from southern Florida to the West Indies, often living amid eelgrass in water 5 to 15 feet deep. This large shell is often used for ornamental purposes. It is also the favorite for making conch chowder and conch steak, those staple foods of the Florida Keys. The shell is often used as a trumpet and occasionally produces semiprecious pink pearls.

The relatively common Atlantic carrier shell (*Xenophora conchyliophora*) demonstrates an alternative way of making itself blend into its environment. Instead of using color for camouflage, this unusual gastropod picks up stones, shell fragments, and bits of coral and then cements these pieces to its own shell. This gives the overall appearance of a pile of old dead shells, providing perfect camouflage and

With intriguing names like shark's eye and cat's eye, the large family of moon shells (Naticidae) has always been a favorite of children.

protection from predators. The added shells also give each carrier shell a broader base, preventing it from sinking into soft mud bottoms of the Atlantic Ocean and Caribbean Sea where it lives.

The Atlantic slipper shell (*Crepidula fornicata*) is commonly found in shallow subtidal waters, sometimes growing on top of each other. The instant you see this shell, you'll know how it got its name. A broad internal shelf protects the creature's soft organs and makes it look exactly like a tiny bedroom slipper.

To my eye, the whelk family of gastropods best illustrates the allure of shell collecting. These classically spiraling shells are widely distributed in cold-water habitats throughout the northern hemisphere. Collectors often confuse the lightning whelk (*Busycon contrarium*) and the pear whelk (*Busycon spiratum*), although close inspection reveals several distinctive differences. The lightning whelk is one of the few left-handed shells, meaning that its shell is coiled in a counterclockwise direction. The shell's common name comes from the well-defined lines that extend down from the knobs of its shoulders, resembling flashes of lightning. On the other hand (figuratively speaking), the pear whelk is right-handed and does not have the pronounced shoulder knobs or distinct markings of the lightning whelk. Both whelk species are commonly found living on sandy bottoms out beyond the surf line.

Tulip shells—with their softly rounded shapes, as opposed to the etched shapes of the whelks—run a close second in classic beauty. The true tulip (*Fasciolaria tulipa*) and banded tulip (*Fasciolaria hunteria*) are both found in bay waters and estuaries. The

Children delight in stuffing their pockets full of these thin, shiny shells (family Anomiidae) and then listening to them jingle as they skip along through the waves.

A popular snack throughout Europe, cockles often are found in abundance on tidal flats.

true tulip exhibits two completely distinct patterns, the solid form and the clouded form, with vivid colors ranging from various shades of tan, gold, and orange to a deep chocolate brown. As its name implies, the banded tulip features narrow, bold, and almost always unbroken bands against a background of blotchy dark "clouds" in a light-colored "sky." Tulips are among the most aggressive gastropods. In fact, banded tulips often eat other tulip species.

The common janthina (*Janthina janthina*), also called the purple snail, survives in one of the most unusual habitats in the molluscan world. Vividly purple in color, they float in large colonies suspended upside down from frothy masses of bubbles in the open ocean. Completely at the mercy of the winds and tides, these snail bubble rafts are often blown into cold waters where they don't survive.

The common periwinkle (*Littorina littorea*) typically lives above water attached to rocks, pilings, or limestone throughout the intertidal zone. Snaillike in appearance and coloration, these gastropods are widely distributed in both cold- and warm-water habitats.

The green keyhole limpet (*Diodora viridula*) features ribbing that radiates out from the apical, centrally located keyhole in its outer shell. Limpets feed on algae that grow on intertidal zone rocks along both North American coastlines. They adhere so tightly to their rocky habitats that they have to be pried off with a knife.

The common nutmeg (*Cancellaria reticulata*) is a resident of coastal shelves, living in water 20 to 30 feet deep. The nutmeg's shell pattern is very consistent, featuring a plaidlike appearance formed by a mixture of white, orange, and brown. The surface of the shell is beautifully textured with crisscrossing beaded ridges.

In appearance, the murex gastropods are the old men of the sea. Their ornately spined shells give them a gnarly, aged look that is quite distinctive from the smoothly streamlined appearance of many other shells. The giant eastern murex (*Muricanthus fulvescens*)

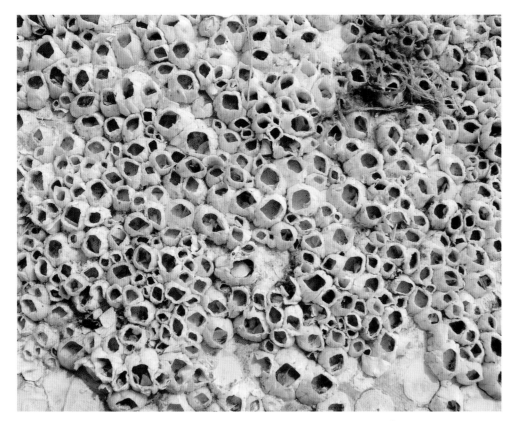

Both barnacles (left) and mussels (below) band together into huge colonies that attach to pilings, seawalls, tide-pools, and any other hard structures within the intertidal zone. *Shutterstock/Carsten Medom Manston, Shutterstock/Dan Mensinger*

A close look at this heart-shaped view of a large cockle shell could reveal the origin of the saying "warm the cockles of my heart."

is commonly found in relatively shallow water, just beyond the breakers in the Gulf Coast. The apple murex (*Phyllonotus pomum*) is also known as the "rock shell" due to both its appearance and solid, hefty construction. Featuring three varices adorned with hollow spines, the lace murex (*Chicoreus dilectus*) has a rustic antique look that sets it apart from every other shell on the beach.

Many gastropods make their "living" by preying on their bivalve kinfolk. These include the Atlantic oyster drill (*Urosalpinx cinerea*) and the Florida rock shell (*Thais haemastoma floridana*). These species feed by using their radulas to bore holes into the shells of their prey and then inserting their prosboci to suck out the soft meat inside. As you might expect, the preferred habitats of these species are intertidal mud flats near oyster beds.

The giant tun (*Tonna galea*) is another species that is a common resident of deep coastal shelf waters. Exhibiting spire whorls separated by deep sutures, this bulbous shell looks like it needs to go on

a diet. The burnished, chestnut-brown color is beautiful, but unspectacular. Because of the amount of time it exists as a free-swimming larval stage, the giant tun is distributed throughout the world.

The lion-paw scallop (*Nodipecten nodosus*) is one of the world's most visually impressive seashells. Perfectly resembling its namesake in shape, texture, and color, this spectacular bivalve is adored and desired by collectors throughout the world. It occurs primarily in deep-water habitats from North Carolina to the West Indies and Brazil.

SHELLS THAT MAKE US WONDER

Many of you share the childhood experience of finding and collecting common jingle shells (*Anomia simplex*) as you skipped along the water's edge. Often you only found pieces of these bivalves with the thin, translucent shells. But collecting them was great fun because of the way they jingled in your pockets or buckets. Successfully finding the baby's foot—the tiny scar where the adductor muscle attached to the shell—always added a special touch to a child's shelling experience. Adult craftsmen also value jingle shells because they're so easily crushed into thin, sparkly flakes.

Moon shells (family Naticidae) go by many common names such as shark's eye (*Neverita duplicata*) and cat's eye and have always enthralled children. Since they live in the intertidal zone, moon shells can readily be found by the careful observer. If you are lucky enough to find a living moon shell, you will be especially impressed by the size of the foot, which when inflated with water is twice the size of the shell!

ANY COCKLE WILL DO

If you've ever wondered about the possible source of the well-known saying, "It will warm the cockles of your heart," look at a cockle shell (family Cardiidae) from the side. You'll see that this bivalve has a distinct heartlike shape.

One of the most widespread families of mollusks, cockle shells are bivalves that are important food items in many parts of the world. In fact, the edible cockle (*Cerastoderma edule*) is a popular delicacy in northwestern Europe. Most cockle species burrow into sand or mud, often occurring in colonies of astronomical numbers in very small areas.

On one of my dad's winter shelling escapes to Marco Island, Florida, he was lucky enough to see a large colony of unidentified bivalves—most likely cockles—in the surf. While strolling along the beach at sunset with his wife, Liz, my dad noticed that the ebbing tide left a shimmering stretch of wet sand, backgrounded by a receding blue-orange wash of gentle waves. In his journal, he wrote about what he saw next: "Then, suddenly as if a curtain had been raised, everywhere we could see on that shimmering background appeared a heavenly dance of creatures, just kind of swaying back and forth with the gentle tide—huge numbers of pink tongues bending with each movement of the tide. And then, as the last vestige of sunlight diminished, they all disappeared," concluding that this was one of the most remarkable sights he had experienced in 25 years of shell collecting.

THE BEAUTY QUEENS

The family of cone shells (Conidae), universally considered to be among the world's most beautiful mollusks, can be found in all tropical and semitropical seas in habitats ranging from subtidal to depths of more than 1,600 feet. There are about 40 Caribbean species and about 600 species worldwide. They feed on live fish, worms, and other mollusks, using a poisonous, harpoon-like tooth. Only Indo-Pacific species have caused human deaths, but Atlantic species can sting. They range in adult size from $^3/_{10}$ of an inch to almost 10 inches and display all the colors of the rainbow combined in every pattern imaginable.

Also among all the world's beautiful shells, wentletraps (family Epitoniidae) are the most prized by collectors. These gastropods with exquisitely spiraling, delicately ridged shells are found on the shorelines of Massachusetts and New York south to Texas and Uruguay. Twenty-four species of wentletraps have been recorded in the waters of North Carolina alone. Arguably the world's most beautiful shell, the precious wentletrap is only found in the Pacific Ocean.

The junonia (*Scaphella junonia*), a member of the volute family, is widely known as the pride of Sanibel Island. Since the junonia, along with all

The junonia (*Scalphella junonia*), a member of the volute family, is the most prized shell on Sanibel Island. If you spot one, dash over and pick it up—before someone else beats you to it! *NHPA/Martin Harvey*

other volutes, resides in deep water, they usually are only found on the beach at the crack of dawn following a stormy night. If you do happen to find a prized junonia, pick it up quickly or it may be collected right out from under your nose. Shelling is a highly competitive endeavor, especially when you're dealing with such a rare species.

The fabulously exquisite chambered nautilus (*Nautilus pompilius*), the only cephalopod with an external shell, is only found from the tropical Western Pacific to the tropical Eastern Indian Ocean (see details on the chambered nautilus in Chapter Two).

THE MONEY MAKERS

Economically, oysters, clams, and mussels are the world's most important mollusks. Oysters are found throughout the globe, living in large beds in tidal zones. The most abundant oyster beds lie along the eastern United States, specifically in the Chesapeake Bay.

Successful oyster farming includes seeding oyster beds, a concept that is strangely similar to the farming practices in the Midwest. While agricultural farmers decide where to plant their seeds for maximum production each spring (consulting the *Farmer's Almanac*, of course), oyster farms continually manage and expand their *spat* settlement seedbeds.

As discussed in Chapter Five, many species of mollusks begin life as tiny larvae floating freely amid the billions upon billions of organisms that form the plankton mass that carpets the surfaces of the ocean. The minuscule percentage of oyster larvae that manage to avoid being dinner for larger sea creatures then become spat and start spiraling downward in search of permanent resting places.

If the descending oyster spat can find clean substrate—like old oyster shells—to attach themselves to, they will begin to grow to harvestable size. If the spat can't find attachment sites, they die.

The entire process in oystering lingo is called *spat set*. If all goes well, in about three years, legal, market-size oysters—exceeding 3 inches in diameter—can be harvested with dredging equipment and sold.

Professional clammers generally have an easier go of it than oyster farmers. Clams burrow into soft sand or mud in the flats of intertidal zones. At low tide, clams conveniently reveal their whereabouts by ejecting spouts of water—like miniature geysers—through their siphon tubes. Then it's just a matter of digging down a few inches with a clam rake and tossing the captured quarries into a wire-frame clam basket designed to let the water out and keep the clams in.

For mussel aficionados, these clinging bivalves are a highly adaptable species, establishing colonies on just about any solid structure. From piers, pilings, and rocky bulkheads, mussels can be found almost anywhere, and there are a number of freshwater species, too. The blue mussel (*Mytilus edulis*) is the world's most commonly farmed mussel, and cultivation takes place in over 20 countries throughout the world. Mussel production and harvesting is closely allied with oyster production. First, mussel beds are created by dispersing spent shells across known production beds. The mussel spat then settle onto the beds where they are harvested by dredges, bagged, and sold at the marketplace.

FAUX MOLLUSKS AND SECOND HOMES

With today's emphasis on conservation and recycling of resources, gastropods should receive an award as the ultimate green species. Each vacated gastropod shell becomes prime potential real estate for hermit crabs, which are the oceanic equivalent of home squatters. Hermit crabs bumble along the ocean bottom with no place to live—until they stumble upon an empty whelk or conch shell.

Hermit crabs must find a new, larger shell each time they moult. Since this may happen several times a year, natural competition for suitably sized gastropod shells may become fierce. Hermit crabs living in shells that are too small are significantly more vulnerable to disease and predation. People who own hermit crabs as pets, and there are many, are well advised to keep a succession of progressively larger

Hermit crabs (*Coenobita compressus*) are opportunistic crustaceans—vagabonds of our ocean floors, always searching for new shell homes to fit their growing bodies. *Shutterstock/Nancy Kennedy*

gastropod shells on hand. After all, who wants to own a crabby hermit crab?

NIXING THE NICHES

Many species of mollusks are very sensitive to water-quality impacts. When pollution becomes severe enough, it destroys the survivability of the ecological niche and the local mollusk populations become either severely reduced or disappear entirely. Two freshwater mollusks, the dwarf wedge mussel and the spiny mussel, were once abundant in river systems throughout the eastern United States. But during the last few decades, poorly controlled development has reduced the populations of both of these species to the point that they are now listed as endangered species by the U.S. Fish and Wildlife Service and fully protected under the 1973 Federal Endangered Species Act.

The ecological threats mollusks face are discussed in detail in Chapter Eight.

The Magnificent Mollusk
Ties to Human History, Past, Present, and Future

Since the time man first walked along the beach, human history has intertwined with the world of mollusks. In many coastal regions today, the clambake is considered a major summertime event, a time for joyous family celebrations toasting the lazy, hazy days of summer. But clambakes have been around for many thousands of years. Only early man considered them to be rights of survival instead of frivolous warm-weather activities.

As you'll read in this chapter, mollusks have served humans well in many other ways. They've provided mankind with essential tools, materials for clothing, artistic designs and inspiration, as well as the earliest forms of currency.

MONEY, MONEY
The oldest and strongest connection between human history and seashells revolves around monetary exchange. By 2000 B.C., money cowries (*Cypraea moneta*) were the established medium of exchange in both Africa and Asia. In fact, cowry shells still hold the record as the most widely circulated currency in the history of mankind. They were the world's primary source of monetary exchange for almost 600 years. They were used into the 1700s in the marketplaces of Burma, India, China, and Africa to trade everything from ivory and rice to rugs and slaves. The international use of shell money finally died out in the late 1800s due to inflation and the increased use of metal coins.

The popularity of cowry shells as currency was due to many factors. First, they were attractive, extremely durable, and uniform in size. From a handling perspective, they were easy to count, weigh, and transport. Most important, they were available in abundant supplies from several secret, secure locations. This made them almost impossible to counterfeit.

Cowry shells also indicated social status. They were placed in the mouths of the deceased at burials. Ancient China had a cowry shell pecking order. At the emperor's burial, his mouth was stuffed full with nine cowries. Feudal lords were buried with seven shells in their mouths, while high officers received five and ordinary officers three. Most commoners' mouths were stuffed with only rice, while those with

Over the centuries, cowry shells (family Cypraeidae*) have been used as standard currency in many parts of the world. *Shutterstock/Yang Xiaofeng*

Modern-day archaeologists investigating long-buried *middens*—ancient dumps—find them full of mollusk shells. *Shutterstock*

some money had a small cowry placed next to the last molar on each side of their mouths. This practice ensured that the dead had plenty to eat and spend in the afterlife.

Among North America's Native American tribes, shells also were used as money. Native American tribes in the Pacific Northwest used the precious tusk shell (*Dentalium pretiosum*) as money for almost 2,500 years. These shells are still avidly collected on Vancouver Island in British Columbia, where they are used primarily for decoration.

The Onondagas tribe of Iroquois Indians of Upstate New York was known as the wampum keepers. White wampum, the central column of a whelk shell, was the customary gift among the tribe, while purple wampum, made from the quahog clam, was given on special occasions. The Onondagas also fashioned wampum beads from bits of whelk and clam shells. They then traded their handmade beads for beaver pelts, canoes, and squaws.

TOOLS AND KITCHEN AIDS

The Calusa Indians of present-day Florida were the original shell collectors in this region. They specialized in adapting shells for use as fishing tools. They perforated clam shells to use as weights for their fishing nets and grooved whelk shells to serve as sinkers for their fishing lines.

North American Indian tribes used shells for every imaginable purpose. Shell jewelry was fashioned to indicate tribal status. Conch shells were inscribed with sacred images and used as drinking cups.

Local beaches also served as a Home Depot for innovative tribal handymen in the Pacific Northwest. They used the sharp edges of clams and scallops as knives and scrapers and the pointed ends of whelk shells as perforators. Thick conch shells made excellent hammers and anvils.

Meanwhile, inside huts and tepees, shells were the all-purpose Tupperware of the day. Large whelk and conch shells served as cups, dippers, pots, and ladles, while giant clam shells made perfect food containers. Modern-day archaeologists investigating long-buried *middens*—ancient dumps—find them full of mollusk shells.

In an interplay of nature and commerce, the scallop has always been the symbol of the Shell Oil Company. In 1892, Shell launched the world's very first tanker, named *The Murex*. The company now has 29 ships named after shells.

THE WORLD'S DELICACIES

Abundance, accessibility, and delectability have made mollusks of all shapes and sizes a primary source of sustenance all over the world. While oysters and clams are best known for feeding human masses, many gastropods—abalones, conchs, and whelks—are also consumed in great quantities.

The common periwinkle is routinely collected and eaten by many people in European countries. Pubs in small villages throughout England still practice the tradition of placing bowls of boiled periwinkles on the bar. Patrons then pick the flesh from the shells with "winkle pins," dunk them in a special sauce, and wash them down with glasses of ale.

Cockles are a popular type of shellfish in both Eastern and Western cooking. They are still collected as they have been for years, by raking them from the sands at low tide. Cockles are boiled or pickled as a snack food in the United Kingdom and are eaten with vinegar. Seafood stalls sell them along with mussels, whelks, and eels. They are also available

THE GREAT SCALLOP SCAM

Many folks who wouldn't even consider eating an oyster or clam—my wife, Debby, being a prime example—consider scallops a favored restaurant entrée. This is because scallops are served as solid chunks of meat only, as opposed to entire bodies of oysters and clams. (Restaurant scallops are described as rounded chunks of the bivalve's large adductor muscles.)

Of course, this raises another issue. Are the chunks of meat on the plate really made from scallops or are they cut out from something else? Restaurants used to economize by cutting plugs from shark fins and stingray wings and substituting them for more expensive scallops. This practice has now been stopped, or so it's claimed. If you're in doubt the next time you order, try this test. Break the meat apart: True scallops break apart easily and all the fibers run in the same direction.

The menu listing can also be a giveaway. If it says something like "giant sea scallops," be wary. Since real scallops are made from only a portion of a relatively small bivalve, it's impossible to serve something "giant." Of course, if you're ordering Coquilles St. Jacques, you're probably in a restaurant where the authenticity of the scallops is not a cause for concern.

pickled in jars and sold in convenient sealed packets (with vinegar) containing a plastic two-pronged fork. Boiled cockles are sold at many hawker centers—open-air markets—in Singapore.

The French word *escargot* is often used on restaurant menus (especially in North America) to refer to snails as a delicacy. In France, escargots are typically only eaten on festive occasions. While not all species of snail are edible, many (116 different species) are. Even among the edible species, the palatability of the flesh

Since they often live clinging to rock above the high-tide line, periwinkles (family *Littorinidae*) can be readily found by the careful observer.
Kevin Adams

varies from species to species. Typically, snails are removed from their shells, gutted, cooked (usually with garlic butter), and then poured back into the shells together with the butter and sauce for serving, often on a plate with several shell-sized depressions. Special snail tongs (for holding the shell) and snail forks (for extracting the meat) are also normally provided.

Because snails eat soil, decayed matter, and a wide variety of leaves, the contents of their stomachs can be toxic to humans. Therefore, before they can be cooked, the snails must first be prepared by purging them of the contents of their digestive system. The process used to accomplish this varies, but generally involves a combination of fasting and purging. The methods most often used can take several days. Farms producing snails for sale in Europe and in the United States typically feed snails a diet of ground cereals.

In the United States alone, there are 60 commercially important clam species. Quahog clams (*Mercenaria mercenaria*) are marketed under a variety of names, including littlenecks and cherrystones. Many other mollusks regularly appear on U.S. menus,

depending on where you are. In the Florida Keys, conch steaks and conch fritters are often the specialty of the house. Steamed and fried mussels are popular in most coastal areas, while calamari—served in a wide variety of international styles—is now an appetizer staple in restaurants across the country.

FASHION STATEMENTS

Shells worn as personal adornments have served many purposes throughout human history, including as aphrodisiacs. The women of Pompeii wore cowry shells to prevent sterility. From ancient to contemporary times, shells also have been used to embellish clothing. For example, often fragile shells, such as angel wings, make excellent sequins for party clothes. Similarly, when strung together in long chains, simple tusk shells make excellent ceremonial headdresses and fancy necklaces. The ancient Greeks used a stylized scallop as a shoulder clasp for their tunics. Scallops were added to the coat of arms of many British families as a reference to Catholic ancestors who had participated in the Crusades.

THE COLOR PURPLE

The color purple originally comes from the world of mollusks. For more than 3,500 years, mollusk shells produced the color purple that distinguished royalty from commoners on the island of Crete.

A species of the murex family (Muricidae) became historically significant in fifteenth century B.C. when the people of Sidon and Tyre discovered a method for extracting a purple dye from the mollusk. The Phoenicians perfected the manufacture of purple dye by extracting it from the spiny dye murex (*Murex brandaris*) and the banded dye murex (*Murex trunculus*). This dye-making process was kept secret for centuries. Only Roman and Byzantine emperors displayed robes colored with the brilliantly hued dye. In fact, Anthony and Cleopatra sailed into the Battle of Actium under sails colored Tyrian purple. In Babylon, idols were clothed in Tyrian purple cloth, while Rome's Emperor Nero was the only person in the empire allowed to wear purple cloth.

Ever since it was used to glorify the appearance of these ancient rulers, the color purple has signified royalty. Burial robes found in tombs of royalty have long been purple in color. In fact, the dye is so long-lasting that the mummy wrappings in museums still show purple colors after thousands of years.

As the status symbol for royalty and power, purple dye eventually became in demand throughout the ancient world. In the sixteenth century in Central America, native tribes poured wide-mouthed purpura snails (*Purpura patula*) into cauldrons and mashed them until a purple ooze started to flow. In 1648, tribe members exported the dye to Spain and created an explosive demand for it. Demonstrating perhaps the first example of resource conservation, the tribesmen implemented measures to prevent over-harvesting of the purpura snails. Instead of collecting and crushing masses of snails, they collected individual snails, gently blew out the dye, and returned the snails to the rocks unharmed.

The ultimate shell status symbol is, of course, the pearl. Pearls are lustrous concretions produced by certain bivalve mollusks, notably oysters. Strangely, these objects of great value to humans start as minor irritations to oysters. When a minute particle, like a grain of sand, gets trapped in an oyster's mantle, the oyster secretes mother-of-pearl around the irritant. Eventually, the mother-of-pearl (or nacre) forms a circular shape. The resulting pearl is composed of aragonite crystals. Pearl oysters are bivalve mollusks of the genus *Pinctada*. All members of this genus share the physiological properties that lead to the production of large pearls of commercial value.

In the 1800s in the United States, it was common for people to collect mussels and look for treasures on the Upper Mississippi River. The treasures they sought were freshwater pearls found in native mussels.

Some historians have compared this treasure hunt to the gold rush in California. The fever to find a pearl was so intense that people literally killed, and then just threw away, millions of mussels. In some areas of the river, entire mussel beds were eliminated. But the effect of the pearl rush on mussel populations was minor compared to the rush to make buttons out of mussel shells.

In 1889, German businessman Johann Böpple started using freshwater mussel shells in America as buttons. Special machines punched out buttons by the millions. The best came from mussels with thick shells, like the yellow sandshell and the pistolgrip. By 1899, 60 button factories were located in the Mississippi River Valley, and these factories harvested over 21,000 tons of mussel shells in the vicinity of Muscatine, Iowa. Less than 10 years after its incep-

Throughout human history, shells have been used to make necklaces, bracelets, and other fashionable accoutrements. *Shutterstock/Chad Littlejohn*

The 1890s brought a huge mussel shell button-making industry to the Mississippi River Valley in the Midwestern United States. *Shutterstock*

tion, the industry supported thousands of workers and was valued at over $25 million.

Mollusks also provided rare finery for the wealthy inhabitants of the ancient Mediterranean area. Tufts of golden silk threads called *byssus* anchored noble pen shells (*Pinna nobilis*) to the sea bottom. These 2-foot-long threads allowed the pen shells to withstand strong currents and underwater swells. With a deep bronze gold coloring, these silky threads soon caught the eye of local fishermen and tradesmen who began to weave them into gloves, stockings, caps, and other specialty clothing.

Byssal fibers were soon in great demand as "silk from the sea." Many believe that the Golden Fleece pursued by the legendary Greek hero Jason and his Argonauts was woven from pen shell byssus threads.

MANKIND'S MOLLUSCAN-INSPIRED MASTERWORKS

For centuries, artisans have used seashells as both inspiration for art and natural canvases for carvings and sculptures. Made famous by the Romans, shell cameos were also extremely popular in France during the Renaissance and in Victorian England. Prime examples include the Lord's Prayer carved in high relief

on a tiger cowry (*Cypraea tigris*) and large motifs carved on emperor helmet (*Cassis madagascariensis*) shells. The helmet shell carvings were often lit from within, providing an extraspecial ethereal glow to these works of art.

As the Aztecs of ancient Mexico depicted their rain god, Tlaloc, rising from a conch shell, many other noted artists have featured mollusks in their work. In his 1509 masterpiece, *The Birth of Venus*, Florentine painter Sandro Botticelli depicts the mythological goddess rising from a scallop shell. Giovanni Bellini's *The Allegory of the Shell* is one of the treasures of the Uffizi Gallery in Florence, Italy. Watercolorist J. C. Xavery used shells to embellish many of her finest works, and moon snails are an integral part of numerous paintings by southwestern American artist Georgia O'Keefe.

Numerous famous photographers have also used seashells to embellish their work. Of particular note, Edward Weston created a series of striking still-life images of shells that are now considered classic works of art. In 1989, his exquisite photo of a polished pearly nautilus sold at Sotheby's for a then world-record price of $115,000.

The meshing of natural and human artistry is perhaps best demonstrated by mother-of-pearl. Produced internally by a variety of mollusks—including the nautilus, abalone, mussel, and trochid shell—mother-of-pearl is a natural symbol of beauty, purity, and nobility. Human artists regularly use bits and pieces of this indescribably iridescent material to create intricate patterns on metal and wooden panels.

Ancient Native American tribes used mother-of-pearl extensively to highlight their traditional native costumes and artworks. The use of mother-of-pearl reached its climax during the Victorian Era when it was used to embellish a multitude of household items, including jewelry boxes, lampshades, and room screens.

Authors have also highlighted mankind's fascination with the natural beauty of mollusks—whether it be Dylan Thomas, *The Cockle Crooks*; John Mawe, *The Shell Collector's Pilot*; William Golding, *The Lord of the Flies*; or Rosamunde Pilcher's 1987 bestseller *The Shell Seekers*. Moon snails even inspired the poetry of Anne Morrow Lindbergh.

COMMUNICATION AND MUSIC

You've all done this as children: Put a large seashell over your ear and listen to the roar of the ocean. Of course, you could have heard the same "ocean's roar" by using a drinking glass, a cereal bowl, or any rounded, enclosed object that fits snugly over an ear. But for many civilizations that existed before Alexander Graham Bell used his new invention to summon "Mr. Watson," seashell communication was serious business.

The conch shell was an essential component of Samurai military communications in the fifteenth through seventeenth centuries. When the enemy was at hand, the army commander instructed the *kaiyaku*, the royal conch shell trumpeter, to blow the signal for an immediate muster. The Greek God Triton, one of Neptune's trumpeters, is often depicted with a large conch shell that he used to summon river deities around their king.

The natural oyster pearl is the icon of the molluscan world's contribution to haute couture. *Shutterstock*

Produced internally by a variety of mollusks, mother-of-pearl is a symbol of beauty, purity, and nobility. *Shutterstock*

MOLLUSCAN TALL TALES: MYTHS, LEGENDS, OR FACTS?

Giant Man-Eating Clams

Between depths of three to 15 feet in the ocean dwell clams that grow more than 4 feet long. As with anything found supersized in nature, the first fisherman to discover these huge bivalves assumed that they ate everything, including drunken sailors who fell overboard. But the truth is that giant clams (*Tridacna gigas*) cannot open their valves more than 6 inches and, as a result, are filter feeders. In fact, they actually farm their own food, consuming symbiotic algae called *Zooxanthella* that grow in great quantities within each clam's fleshy mantle lobe.

The Whistling Oyster

In his 2005 book, *Out of My Shell*, S. Peter Dance provides the following tale: One night in 1840 after closing time, a pub owner in London heard distinct whistling sounds coming from a tub of oysters. Upon further investigation, he found that a single oyster was producing the eerie sounds by pushing air through some tiny holes in its shell. Being a shrewd businessman, the pub proprietor immediately dubbed his establishment the Whistling Oyster Tavern.

Word spread quickly, and the Whistling Oyster was soon the talk of London. Among the flock of new customers was a Yank from New England. Upon hearing the oyster's whistle, he immediately proclaimed that it was nothing compared to an oyster he knew in Massachusetts that "whistled 'Yankee Doodle' right through and followed its master about the house like a dog."

Snap-Happy Oysters

Over the years, numerous reports have surfaced about oysters being kept live in people's kitchens and pantries. While that may seem odd, what's even odder is how these shells have served as a better mousetrap. At night when these bivalves open their shells, hungry mice stop by for a snack. Then the oyster valves snap shut, and the mice are trapped headfirst with their tails sticking out. According to author S. Peter Dance, as many as three mice have been caught at one time by a single oyster. This is still presumed to be the world record for number of mice caught by a single mollusk.

Clash of the Titans

In October 1966, two lighthouse keepers at Danger Point, South Africa, watched as a giant squid attacked a baby southern right whale. As the whale's mother watched helplessly for almost two hours, the monster squid repeatedly dragged the baby whale under trying to drown it. Fighting valiantly, the little whale would stay under for 10 to 12 minutes, then resurface for a blow or two before being dragged down again by the squid's powerful arms. The squid finally won the battle and the baby whale was never seen again.

Since we humans tend to favor our own kind—mammals, including intelligent docile creatures like whales—most of us probably hate the beastly giant squid that did this dastardly deed. Well, take heart *cetacean* (marine mammal) lovers! It just so happens that giant squid are the primary—maybe only—item on the menu of the sperm whale. Feeding sperm whales routinely dive below 3,000 feet. They have sharp, conical teeth that they use to snap up prey, chiefly medium- and large-sized squid, at these depths.

So, our deep ocean trenches serve as the arenas for what must certainly be one of the most intensely ferocious wildlife struggles on earth. In fact, sperm whales are so dependent on giant squid for food that researchers radio-track the whales to determine locations and migratory patterns of the squid.

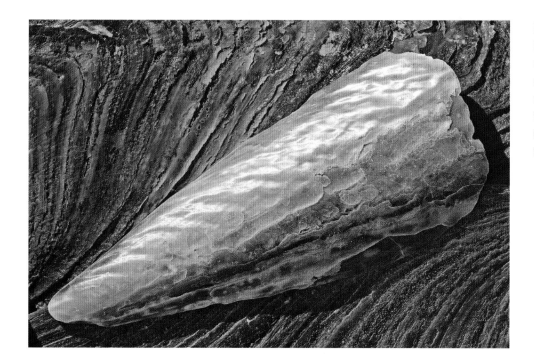

Because most species live in soft sand or mud, pen shells (family Pinnidae) rely on lengths of byssus—tufts of strong silk threads—to hold them in place.

Decoratively engraved and hand-colored, shells were also adapted as musical instruments. Conch shell trumpets were among the world's first wind instruments. They were used to promote meditation in Japan, announce the arrival of chieftains in Polynesia, and summon believers to prayer in India.

By the end of the twentieth century, shells were recognized as musical instruments in their own right. Harrison Birdwhistle's opera, *The Mask of Orpheus*, includes a series of great blasts from a collection of conch shells, while the San Francisco Bay Area's jazzman Steve Turre is widely known for his mastery of conch shell trumpets. In 1993, he produced a ground-breaking CD entitled *Sanctified Shells*. And according to an article in the April 6, 1996, edition of the *London Daily Telegraph*, "Shellfish are very much on the menu for the BBC Symphony Orchestra this week."

BUILDINGS AND ARCHITECTURE

Seashells have influenced the world of architecture from the time of the ancient Romans. The world-famous Trevi Fountain in Rome is symbolic for the triumph of Oceanus riding on a seashell pulled by one calm horse and one wild horse. Many European buildings, including the works of Da Vinci and Gaudi, have also been inspired by the natural beauty of shells. Theodore Cook's books, *Spirals in Art and Nature* and *The Curves of Life*, featured buildings with distinctive

Helmet shells (family Cassidae) were used by natives on South Pacific islands as cooking utensils.
Shutterstock/Darryl Brooks

Triton's trumpet *(Charonia variegata)* is a beautifully decorated shell that played an important role in Greek mythology. *Shutterstock/Galena Barskaya*

spiral staircases. Cook's featured designs were especially influenced by the perfectly symmetrical shapes of the precious and rare wentletrap.

For architectural ornamentation, no shell is more popular than the scallop. An exquisite example of nature's graceful simplicity and perfect design, the scallop's image has been borrowed for niches, facades, tombs, and pedestals. Some famous examples are the Queen Ann Shell Porch, Bernini's *Fontana delle Api* or Bee Fountain, and the St. James Scallop. In seventeenth-century Paris, shell-encrusted rooms provided a perfect demonstration of the Rococo art form.

Among the world's most famous shell-designed buildings are the Van Wezel Performing Arts Hall in Sarasota, Florida, and the Sydney (Australia) Opera House, which was inspired by the comb oyster (*Lopha*

cristagalli) and cemented as a cultural icon during the 2000 Summer Olympics. The Frank Lloyd Wright–designed Guggenheim Museum in New York City mimics the shape of the miraculous thatcheria shell.

FOUNDATIONS AND STRUCTURES

Architects and engineers have long marveled over the strength and durability of mollusks. In his delightful 2005 book *Out of My Shell*, S. Peter Dance describes a mussel-bound bridge in Bideford, England. According to Dance, an 1831 letter explains how live mussel colonies provided the solution to a design problem that had city engineers scratching their heads for years. Through test after test, no one could come up with a construction material strong enough to build bridge arches that could withstand the extremely powerful local tidal currents. Then someone in the city government pointed out that colonies of the common blue mussel were thriving on rocks where the bridge was to be constructed. Sure enough, the mussels produced strong byssal threads that adhered to the stonework of the bridge and the project was finally completed.

Oyster and clam shells are commonly crushed and used to make driveways, highway beds, railroad beds, and even foundations for cities. Prime examples are portions of the City of Mobile, Alabama, and a 6-mile-long road from downtown New Orleans, Louisiana, to Lake Pontchartrain, both of which are built on foundations of shells from the common rangia clam (*Rangia cuneata*).

PHARMACEUTICALS AND MEDICAL RESEARCH

During the 1800s, before the advent of "modern medicine," mollusks were used to treat an array of ailments. Ground to a fine powder, cockle shells were believed to be good for the heart, while crushed snail shells were administered to treat everything from bad colds to consumption (now known as tuberculosis). Even pearls were ground up and used to treat stomach ailments.

Throughout history, architects have been inspired by the designs of seashells. The results of their inspiration can be seen in works from big—the Sydney Opera House in Australia—to small—single-family beach homes.
Shutterstock

Everyone, from young children to retired senior citizens, enjoys spending time walking the beach and collecting shells on Florida's Sanibel Island.

In the 1960s, researchers discovered that extracts from hard-shelled clams called mercenine were strong growth inhibitors of cancer cells in mice. They also found that a substance in raw abalone juice, called paolin, was effective against penicillin-resistant bacteria.

Today, mussels are frequently used as sources of compounds for all types of medical research projects, particularly those focusing on developing anticancer and antiviral drugs.

PLEASURE AND PASTIMES

From a recreational angle, shell collecting is one of the most popular pastimes in the United States. Almost every state—coastal and inland—has numerous shell clubs that are part of the Conchologists of America (COA). COA is an international organization of pro-

fessionals and amateurs interested in the beauty of shells, their scientific aspects, and the collecting and preservation of mollusks. Members include novice, as well as advanced, collectors, scientists, and shell dealers from around the world. Local club activities include monthly meetings with featured speakers, weekend beachcombing trips, regional shell shows, and an annual convention.

Finally, if you're really interested in getting into conchology, take a trip to Sanibel Island located near Fort Myers on the southwest coast of Florida. Sanibel is known far and wide for its fabulous shell-collecting opportunities; plus, one of the best shell museums is located there. Spend a day at the Bailey-Matthews Shell Museum and you'll learn more about mollusks than you ever thought possible!

A Shell for Every State

If you live in a coastal state, you may have an official state mollusk. North Carolina was the first state to hold this distinction, choosing the Scotch bonnet (*Phalium granulatum*) in 1965. Since then, 13 other states have adopted state shells, several of them promoted by shell clubs.

State shells that were adopted largely for their beauty include the following gastropods: Massachusetts' New England neptune, North Carolina's Scotch bonnet, South Carolina's lettered olive, Georgia's Kiener's whelk, Florida's horse conch, Alabama's Johnstone's junonia, Texas' lightning whelk, and Oregon's Oregon triton. Bivalves adopted as state shells include the eastern oyster (adopted by Connecticut, Virginia, and Mississippi), northern quahog (Rhode Island and New Jersey), and Atlantic bay scallop (New York).

The state that picked the state shell with the most appropriate name is Oregon, which adopted the Oregon triton. Two official shells have scientific names commemorating the commonwealth of Virginia: *Crassostrea virginica* and *Chesapecten jeffersonius*. The first is Virginia's state shell, the eastern oyster. The second is Virginia's state fossil. Its name recalls Chesapeake Bay, which Virginia shares with Maryland. But *jeffersonius* stands for Thomas Jefferson, perhaps the most famous of all Virginians.

The northern quahog makes sense for the two states that adopted it. This bivalve ranges from Newfoundland south to Cape Hatteras. But *quahog* is a Native American name that is often associated with New England and neighboring coastal states. Another celebrity clam with a Native American name is Washington State's geoduck (pronounced "gooey duck"). But it's not an official state symbol—yet.

While New Jersey is sometimes called the Clam State, it is better known as the Garden State. Maryland was once called the Oyster State, but now it's better known as the Bay State.

Perhaps the most prominent shell nickname is the Conch Republic. That's what some people call a group of islands known as the Florida Keys. Residents had long been known as Conchs when they became angry at the federal government and declared their independence. As of 2007, the Keys are still a part of the United States. And most of the conchs that tourists buy there are imported from the Caribbean.

The largest state shell is Florida's horse conch. It can grow nearly 2 feet long. The horse conch may live as long as 12 to 14 years, which also makes it the oldest state mollusk.

The horse conch (*Pleuroploca gigantea*) holds several distinctions in Florida. As the state's largest and possibly oldest mollusk, it was chosen as the state shell.

Chapter 8

Oceanic "Canaries"
ENVIRONMENTAL THREATS, ISSUES, AND SOLUTIONS

One of the world's most dangerous jobs is being a coal miner, where being trapped by explosions, flooding, and cave-ins is always a threat, as is developing black lung disease. For centuries, miners took caged canaries down into the pits with them, not to hear them sing, but to provide early warning signs of every miner's nightmare: the presence of carbon monoxide gas. When the canary died, the miners knew they had only a precious few seconds to get to the surface before they ran out of oxygen.

Because of their extreme sensitivity to changes in water quality, many mollusks serve as "canaries" for the world's oceans. When a mollusk population shows a steep decline over a short period of time, often a year or less, it's a sure bet that something has gone terribly wrong in their oceanic habitat. Too often, the loss of shellfish beds is lamented as just another economic hardship, without realizing that the problem extends much further than the loss of local income for fishermen and their families.

Fixing a mine that has been contaminated with carbon monoxide gas is a relatively simple process. While it may take a few days, mine inspection crews identify the source of the leak, correct the problem, and let the miners go back to work in a safe environment. Unfortunately, fixing the catastrophic loss of shellfish beds is not a simple process and, in some cases, the answers are never found. In fact, more often than not, the problem continues to grow, eventually resulting in widespread, permanent damage to an entire marine ecosystem.

While this scenario may sound overstated, it is all too often true. To prove it, you need to look no further than the Chesapeake Bay, once the most productive estuarine-oceanic ecosystem in the world. It became a virtual marine desert in a few short decades. Annual harvests of the Atlantic—or native—oyster (*Crassostrea virginica*) in Virginia and Maryland have declined by 97 percent from 6,000,000 bushels in 1960 to 150,000 bushels in 2006.

Farther south, North Carolina has lost 12,000 acres of shellfishing beds in the past 20 years. Another 56,000 acres of shellfishing beds along the North Carolina coast are permanently closed because of pollution, disease, overfishing, and loss of oyster reefs. The state's oyster harvests have plummeted from 1.8 million bushels in the early twentieth century to only 50,000 bushels in 2007. As oyster stocks have been decimated, the livelihoods of entire communities have been destroyed.

Because of their extreme sensitivity to changes in water quality, seashells function as the "canaries" of the oceans. *Shutterstock/Alex James Bramwell*

Oyster harvests in the Chesapeake Bay declined by 97 percent in less than 50 years. *Shutterstock/Darryl Sleath*

Marine mollusks are both extremely sensitive and sessile, meaning that they can't move to a new location when their habitat begins to deteriorate. (Oysters in particular are very significant in water pollution control. Each oyster filters or cleanses 50 gallons of water a day. A single oyster can process 3.5 gallons of water per hour through its siphon tubes.) When a marine mollusk population exhibits a severe decline in species diversity, total abundance, or both, it's time for concerned citizens to sit up, take notice, and do something. But before detailing ways you can help improve the declining health of the world's oceans, let's examine some of the reasons these waters have become threatened.

THE PROBLEMS

Poor Land Development

Throughout most of the 1900s, the United States' coastal land development practices were abysmal. The worst mistake has been the development of barrier islands and ocean fronts. Barrier islands actually move around in response to oceanic storm cycles.

The "sand" on Sanibel Island's famous beaches in Florida actually consists of pulverized pieces of seashells. *Shutterstock/ShutterVision*

Because of this, they provide barriers or buffers to a storm surge's impact on the mainland. If they are left undisturbed, barrier islands provide significant protection for mainland development.

Unfortunately, developers have taken advantage of the public's desire for oceanfront properties. Seaside mansions, roads, and supporting retail establishments now cover many barrier islands. In effect, this amounts to real estate suicide, since all barrier island developments are doomed to fail at some point. No matter how much money is spent trying to stabilize oceanfront developments with seawalls and bulkheads, the ocean's power will eventually win out and cause the catastrophic loss of property and lives. It's not a question of if but when a barrier island development will be obliterated by a hurricane or powerful nor'easter.

In the introduction, I talked about my experiences growing up with summers spent on Sandbridge Beach in Virginia. Although my Sandbridge memories highlight my childhood, the truth is the development of this barrier beach, situated on a strip of land between the Atlantic Ocean and Back Bay, should have never been allowed. Year after year, oceanfront mansions are pummeled by winter nor'easters. During the early 1980s, the wealthy property owners insisted that someone had to do something to protect their homes. In response, the federal government approved construction of 5 miles of bulkheads, consisting of 40-foot-long steel girders pounded 20 feet deep into the high-tide zone and guy-wired back into the foredunes. Property owners paid $50,000 each for this work. In many cases, the bulkheads didn't even make it through the first winter's storms. The storm surges simply overtopped the bulkheads and mashed them flat from the inside out. Today, little remains of this attempt to control the ocean's power.

Not only did Sandbridge residents lose their money, but they also lost their beautiful wide beaches. While temporarily stabilizing Sandbridge, the bulkheads caused a significant change in the local coastal flow dynamics. Storms began eroding sand from the Sandbridge area and depositing it farther south, soon reducing Sandbridge's beaches to slivers. Now each spring, Sandbridge residents pay to have sand dredged up and pumped in to create new beaches for each summer recreation season.

Hundreds of other examples of catastrophically poor judgment in oceanfront developments—from Bar Harbor, Maine, to Brownsville, Texas—can be cited. They all have had massive impacts on the

Generally, sea walls don't provide permanent protection for expensive waterfront mansions. A better approach is to build farther back from the water and leave natural coastal resources undisturbed.

coastal flow dynamics, causing changes in erosion-deposition patterns and cycles that bury oyster beds and obliterate clam flats.

Any oceanfront development takes a toll on both the human and natural environments. The Mississippi Gulf Coast provides a prime example. Eighteen months after I was born in 1947, my family was forced out of our home in New Orleans, Louisiana, by an unnamed hurricane that wreaked havoc on the coastline from Mobile, Alabama, to New Orleans. In 1969, I drove through this same area and witnessed the total devastation—resembling a bombed-out city—wrought by Hurricane Camille.

And then there was Hurricane Katrina in August 2005. While New Orleans received most of the national attention, much of the Mississippi Gulf Coast was again leveled by hurricane winds and storm surges. In another example of what can happen to a barrier island during a hurricane, that same year Hurricane Rita completely obliterated Paradise Island, Alabama. That makes three total wipeouts of the same coastal area over a period of just 60 years. Yet, less than a week after Katrina hit, government officials were already talking about building the area back again, this time bigger and better. Obviously, someone is not paying attention to nature.

Industrial Pollution

For centuries, the industrial world believed it couldn't pollute the ocean—there was just too much water there to do harm. Consequently, the oceans were treated as huge hazardous waste dumps. Materials that were considered potentially harmful were loaded onto barges, hauled offshore a few miles, and dumped—untreated—into the water. Fortunately, the U.S. Environmental Protection Agency (EPA), National Marine Fisheries Service (NMFS), and other federal/state regulatory agencies have largely curtailed this practice. But has this regulatory control been implemented too late?

Overharvesting

For centuries, oyster and clam beds have been called the gardens of the sea. Today, widespread habitat destruction, exemplified by the use of draggers that "clear cut" the ocean floor, have impacted all the ocean bottom's resources, creating underwater deserts in the process. Decimation of oyster stocks in the Chesapeake Bay is partly due to the use of modern harvesting equipment such as draggers. These high-tech marine vessels created short-term bonanzas of revenue for the bay's oystermen. But by the time they realized the damage that was caused by the draggers, it was too late.

For her sophomore thesis in the spring of 1998, my older daughter, Mariah—a biologist and avid naturalist—studied the fishing techniques in the tiny fishing village of Puerto San Carlos in Baja California Sur, Mexico. A decline in annual harvests—especially of shrimp—was causing regional concern. If the shrimp disappeared, so did the livelihoods of most of the families in and around Puerto San Carlos.

It didn't take long for my daughter's team of research scientists to figure out the problem. The trawling nets used by the village fishermen were simply too efficient. They were like giant vacuum cleaners, scouring the bottom of the ocean. Everything they passed over ended up in the nets.

Back at the docks, the fishermen off-loaded the shrimp and other marketable species and hauled them off to nearby cities. The rest of the sea life—often the majority of the catch—went to waste.

Heavy foot traffic on sand dunes causes the loss of stabilizing dune grass, which in turn exposes the dunes to severe wind erosion. *Shuttershock/Liz Van Steenburgh*

Among the species that ended up dying for nothing—often because these animals couldn't handle the change in water pressure—were bottom-dwelling puffer fish, sea cucumbers, starfish, and crabs.

The Puerto San Carlos situation is not uncommon. Technological advances in fishing gear often result in an astonishing increase in annual harvest—for a while, that is. Because the trawling nets caught everything in Puerto San Carlos, the number of breeding adult shrimp decreased each year. As soon as the number of shrimp being caught exceeded the number of shrimp being produced, the annual harvests started to decline.

The solution in Puerto San Carlos—and everywhere—is blending a healthy dose of common sense into each fishery management program. That can be achieved by setting definite harvest quotas and seasonal limits designed to maintain healthy breeding populations of target species. Increasing profits in fishing operations—both big and small—doesn't require working harder, just working smarter!

Stormwater Runoff

Uncontrolled stormwater runoff—otherwise known as nonpoint pollution—is another big problem in U.S. watersheds and river basins. When a piece of land is developed, much of the natural vegetation is removed and replaced with buildings, parking lots, roads, and sidewalks. This results in the loss of the natural absorption and filtration of contaminants provided by forests, fields, and meadows. Instead of replenishing aquifers and freshening rivers, rain falling on the site now transports everything from oil and gas in parking lots to fertilizers and insecticides on lawns to the nearest streams. These streams then flow into rivers, the rivers into larger rivers, eventually reaching estuaries that feed the oceans.

Fortunately, federal, state, and local governments now recognize the incredible amount of untreated pollution that enters this country's natural drainage systems every time it rains. Most require developers to include stormwater management systems in their site design plans. Unfortunately, these stormwater management systems are designed on a site-by-site basis with no integrated plan for managing runoff on a regional or watershed basis. So while each new development site may now treat runoff before it's discharged from the site, there are still many problems to be solved in properly sequencing runoff, especially in rapidly developing watersheds.

Consequently, runoff from watersheds that were 95 percent forests and farm fields 10 years ago but are now covered by shopping malls, office-industrial parks, and residential subdivisions has increased dramatically. New stormwater runoff regulations are being written to control and protect both the quality and quantity of increased runoff from developing areas. But while these regulations are well intentioned, they are not doing the job that needs to be done. I'll talk about how this situation can be vastly improved later in this chapter.

Overcollecting

Overcollecting, especially of living mollusks, has become a potential problem, especially in such shelling hot spots as Sanibel Island, Florida. To contain and eliminate this problem, Sanibel has recently implemented complete bans on the taking of shells that contain live mollusks.

THE SOLUTIONS

Improve Land Development Practices

One of the best ways to improve land development practices on coastlines is to use low-impact development (LID) and leadership in energy and environmental design (LEED) sustainable design practices. Instead of the standard practice of designing a development and then making the site fit the design, LID and LEED start by assessing the natural features of a piece of property, then making the development fit those natural features. In doing so, developers work with the natural contours and drainage patterns of the land, minimizing the amount of cut-and-fill work that has to be done. The result is a development that minimizes the loss of natural resources and is much more pleasing to the eye.

Best of all, LID and LEED design practices actually save developers lots of money by significantly reducing the amount of earthwork that has to be done. Then, once the project is finished, they realize greater income in the selling price of the lots due to the amenities provided by protecting natural resources and maintaining the visual aesthetics of the natural landscape. Fortunately, many developers and regulators now realize LID and LEED's benefits, and the new design technologies are quickly catching on around the country.

Federal agencies and many states (including Massachusetts, Maryland, Washington, and Minnesota) are in the forefront in researching and promoting the benefits of LID and LEED. But as with many government initiatives, especially those that deal with big business, these changes are slow to come. Any sweeping regulatory initiatives like this often meet stiff opposition from development-oriented lobbyists. As a private citizen concerned about the long-term health of the oceans and the planet in general, you can help by writing to your congressmen and congresswomen in support of LID and LEED. If you're interested in find-ing out more, the internet is rife with detailed information about these new desgin technologies. Just go to your favorite search engine and enter "low impact development," and you'll find a wide selection of sites addressing everything from rain gardens to porous pavements and green rooftops.

Control Stormwater Runoff

The most critical component of LID is controlling stormwater runoff at newly developed sites. Due to the implementation of federal regulations such as the National Pollution Discharge Elimination System (NPDES), the United States has come a long way from the days when sites were developed without regard for the erosion and pollution caused by stripping land of its natural vegetation. Now, every site plan must include detailed erosion and sedimentation control plans. But when it comes to treating site runoff for pollution removal and avoiding increased runoff volumes, this country still has a long way to go.

The typical approach to stormwater management is known as "pipe-and-pond." Through a system of curbs, gutters, and catch basins, all site runoff is directed to one huge detention basin, often an acre or larger. You've probably seen these basins next to large, new developments. They're lined with rip-rap

No amount of money or engineering design can control the power of the oceans. The long-term answer is better planning and control of beachfront development.

As shown in this photo, sand dune restoration is one of the best things we can do to protect our oceanfront resources. Houses should be built behind—instead of on top of—fore dunes. This allows the fore dunes to do their job of buffering inland development from the fury of storms. *Shutterstock/Justyna Furmanczyk*

and, over time, become community eyesores, filling up with foul-smelling water and trash.

Low-impact development presents a much better option for stormwater management. It involves the use of best management practices (BMPs), such as vegetated swales and small groundwater infiltration chambers scattered throughout the development. The design is based on dividing the development area into subwatersheds and then using small BMPs to receive and treat runoff immediately at the points where it's generated. A prime example is designing parking lot islands as minor swales planted with native vegetation. The parking lot is then graded so that each parking area drains to one of these vegetated swales. The swales receive the runoff water that first feeds the planted vegetation and then infiltrates back into the ground through crushed stone that underlies the planting substrate in each swale. This LID design produces multiple benefits. All stormwater runoff is contained within the parking lot, the native vegetation provides habitat for a variety of songbirds, and fresh water—filtered and cleansed by percolating through the crushed stones—is recharged to the underlying groundwater.

Other BMPs that fit well within any LID design scheme include rain gardens, porous pavements, infiltration trenches, and enhanced buffer zones.

Rain gardens are pockets of planted native vegetation that collect localized runoff, often from rooftops. They then use the planted vegetation to absorb contaminants and, as with the swales, use an underlying rocky substrate to further cleanse and discharge the water back into the ground. As with all BMPs, the beauty of rain gardens is that they can be added to a site design wherever there is room, fitting a space that may be as small as one-tenth of an acre.

Porous pavements are ideal for low-traffic areas, such as auxiliary parking lots or driveways for delivery trucks. Unlike standard asphalt paving, porous pavements have lots of void spaces that allow rainfall to infiltrate back into the groundwater instead of running off and having to be treated at other locations on the site. Infiltration trenches are designed to surround a parking lot. The parking lot is pitched so that rainfall runs toward the edges where it is collected by the trenches and then recharged to local groundwater.

Enhanced buffer zones are areas along the sides and rear of retail developments and office parks. Upland native vegetation is planted in these areas, providing additional absorption and treatment of runoff contaminants. In office park settings, these buffer zones often include walking paths, bird feeding stations, and bird nesting boxes.

Because they provide abundant hiding and feeding places for young fish and shellfish, salt marshes are known as the world's nurseries.

In terms of environmental damage, dragging shellfish off the ocean's bottom is analogous to clear-cutting a forest.

Many companies are adding a green roof component to their building design scheme. A green roof is created by placing a layer of topsoil planted with drought-tolerant native vegetation on top of leak-proof membranes on a roof's surface. Green roof BMPs have numerous advantages—rooftop gardens and parks, decreased heating and cooling costs, and enhanced air quality—which make them too lucrative to pass up.

Control Harvesting

Controlling the harvest to protect fishery stocks for the long term would seem to be a no-brainer. Unfortunately, this isn't the case in many situations. For example, Maine's Monhegan Island, situated 10 miles out into the open Atlantic and accessible only by ferry boats, has long supported a community of lobstermen who are also shrewd conservationists. They protected the lobster stocks in their 3-mile territory around the island by trapping for only six months a year, from January 1 to July 1. The rest of the year they allowed the lobster population to recover and reproduce for next year's catch.

Meanwhile, the mainland communities around Monhegan Island lobstered year-round and soon found that their stocks were severely depleted. In desperation, often under the cover of darkness, they set their traps in the waters belonging to the Monhegan Islanders. In no time, the Monhegan situation turned into an all-out lobster war, which lasted for several years. The state legislature finally stepped in and passed regulations protecting the Monhegan lobstermen and their right to regulate their own fishing waters. Lobstermen who didn't respect these rights faced loss of licenses, severe fines, and jail time.

While most coastal states have regulations governing the harvest of commercial fisheries, the ultimate solution still depends on the honesty and integrity of the people involved in the industry. In many cases, this comes down to mandating that fisherman attend educational workshops on managing and conserving their resources before they are issued licenses.

Get Involved!

Many of the major watersheds in coastal states have established associations of private citizens that act as antipollution watchdogs. They attend hearings and present testimony to ensure that proposed developments include necessary safeguards to protect the tributary streams throughout the watershed. Having coordinated public hearings for hundreds of proposed developments, I can tell you that these watershed associations have plenty of clout with local planning and zoning and wetland commissions.

If your local community has a watershed association, join it. If you don't have one, take the initiative and start one. You can make a huge difference in the way things are done in your community and your own backyard.

In 2003, scientists, fishermen, policymakers, and educators joined forces with the North Carolina Coastal Federation (NCCF) in preparing the *Oyster Restoration and Protection Plan: A Blueprint for Action*. During the past four years, members of the plan's steering committee and hundreds of private citizens have worked to make the plan an overwhelming success. Spurred by annual legislative oyster roasts, community leaders are recycling "used" oyster shells from restaurants and private homes back into known oyster seedbed areas. In the process of restoring oyster reefs all along the coast, the oyster plan is also leading citizen support for a comprehensive coastal restoration and protection strategy. For details on the plan or the NCCF, go to www.nccoast.org.

If you're a writer or a photographer, use your skills to document what's wrong and become a champion for cleaning things up. Take your lead from icons such as Anne Morrow Lindbergh, who worked tirelessly to promote conchology and resource conservation and then documented her efforts in her landmark book, *A Gift from the Sea*.

Join Local Shell Clubs

The Conchologists of America, Inc. (COA) is a national organization established for the benefit of both amateur shell collectors and professional malacologists. Each coastal state has a local chapter that is always interested in attracting new members. You'll find that joining a shell club in your state will not only keep you abreast of all the latest happenings in the molluscan world, but will offer monthly meetings and field trips that will give you new conservation-minded friends—as well as a hobby you can enjoy for the rest of your life.

Chapter 9

Shell Collecting 101
Tips and Top Hunting Sites

Picture this: It's a perfect day with the temperature in the mid-80s. A few puffy clouds are wafting around in a crystal blue sky, and a gentle breeze is tickling your earlobes. You're standing ankle-deep in cool water with waves crashing across your shins. Your worries at work and home have melted away; you are lost in total concentration on your newfound pastime: finding nature's hidden jewels at your feet.

I hope this book has inspired you to bring this vision to reality. If so, you'll need to know a few things to prepare for your first venture into the realm of conchology.

HOW TO PREPARE

Before you head off to your selected destination, make sure you do the following:

- Check with the local chamber of commerce or tourist bureau, either on the Internet or by mail, for recommendations on local lodging and restaurants.
- Plan your trip carefully. For my money, you can't beat Triple A's Trip Tix for this. AAA is also very helpful with lodging recommendations and reservations, and you usually get a discount to boot! Then buy a good travel guide and you'll really be set.

- If you're planning a trip to a national or state park, be sure to check with its visitor's center about the best locations, dates, and times for finding mollusks.
- Check the weather forecast for your destination and travel period and then pack accordingly. Even if you're headed to a subtropical climate, throw in a windbreaker, sweatshirt, hat, extra socks, and backup footwear. If you've ever been caught in a tropical downpour, you know that everything—including the inside of your shoes—gets soaked.
- Speaking of shoes, make sure to take along proper footwear for walking around slippery tidepools or through squishy tidal flats. I always take a pair of aqua shoes—shoes with elastic, quick-dry tops and treaded soles—backed up by old running shoes and Tevas for my relaxation periods and other nonbeach time.
- Make sure your travel list includes safety and comfort necessities such as insect repellent, sunscreen, a water bottle, sunglasses, rain gear, a flashlight, and a first aid kit.
- Items you'll find useful in your collecting efforts include a 7x10-inch collecting net (you

Mussels, barnacles, and multicolored sea stars (class Asteroidea) are the featured residents of this tidepool along the Oregon Coast. *Shutterstock/Andy Piatt*

can pick one up at your local aquarium store); a heavy-duty, 10-gallon-size plastic bag; 10 one-gallon-size Ziploc bags; 20 sandwich-size Ziploc bags; five hard plastic containers (variety of sizes—Tupperware will work); paper towels; shell field guide (the Peterson series is always good); a log book (for taking notes); and a backpack for hauling everything, including shells, around. (The uses for all this gear are explained later in this chapter.)

WHERE TO GO

The East and Gulf Coasts

Without question, North America's best shelling beaches are found on the Atlantic and Gulf Coasts of the United States. From Maine to Texas, these coastlines are dotted with great conchological getaways. Best of all, most of these beaches are located on land that is protected by either federal (the National Park Service or U.S. Fish and Wildlife Service) or state government agencies. From north to south, here are some of the best shell-collecting sites:

Maine

Nicknamed the "Pine Tree State" for its 17 million acres of forestland, Maine also features an unbelievable 3,478 miles of coastline. That's equivalent to the dis-

Once you begin collecting, always check on the local rules about taking material away from a beach. Most beaches prevent taking live mollusks, and some also limit the number of shells you can collect each day. *Shutterstock*

tance you would fly if you traveled from Portland, Maine, to San Francisco, California! How can this be possible? The state's eastern coast consists of southeast-oriented peninsulas, seemingly located every 15 miles as you drive north along U.S. Highway 1. Each peninsula extends out from the mainland for approximately 15 to 20 miles, accounting for a total of 30 to 40 miles (counting both sides) of coastline for each peninsula. So if you do the math, you'll find that Maine's nearly 3,500 miles of coastline are a reality.

If you're a fan of sandy beaches, you won't find many in Maine. Ninety-eight percent of the Maine

Snails decorate the bottom of a tidepool in Acadia National Park, Maine.

Barnacles hold fast to multicolored rocks along the shores of Monhegan Island, Maine.

coast is rock-bound, featuring reddish-brown granite boulders, topped with emerald green evergreen forests. Also, if you like to swim, Maine's not for you. The temperature of the ocean seldom exceeds 60 degrees Fahrenheit, even in August!

Since sandy beaches are so rare in Maine, your best bets for shelling are the tidepools that pockmark the rocky coastlines. Filled with crystal-clear water, Maine tidepools are like dazzling abstract three-dimensional works of art. Best of all, the bottom of each pool is literally crawling with life, including green crabs, jellyfish, and a host of mollusks. Mollusks commonly found in Maine include the soft shell clam, common periwinkle, blue mussel, Atlantic deep-sea scallop, New England whelk, Atlantic plate limpet, Atlantic slipper shell, Atlantic jackknife clam, Iceland scallop, eastern mud whelk, common northern whelk, and northern rough periwinkle.

From north to south, Maine has a wonderful array of tidepool-filled natural areas, including the following:

West Quoddy Head State Park is the easternmost point in the continental United States. Standing on 100- to 200-foot-high cliffs, visitors can see the surf crashing on Canada's Grand Manan Island. West Quoddy is also known for its candy-striped lighthouse.

Covering 40,000 spectacular acres, **Acadia National Park** is the oldest national park east of the Mississippi. With an interior intertwined by historic carriage roads for hiking and biking, the park has a coastline that supports both pockets of sandy beach and oodles of tidepools. Sand Beach, Acadia's largest sandy beach, is located on Mount Desert Island in the park. Sand Beach features broad inlet streams and shallow intertidal ponds that are excellent places to search for mollusks. Of course, tidepools can be found just about anywhere you stop along the park's outer loop road. The blue mussel beds at the park's Schoodic Point should not be missed.

Monhegan Island is a two-hour ferry ride from the peninsula towns of Port Clyde or Boothbay Harbor. Once there, you'll feel like you've been transported 60 years back in time. Except for the rusted pickups of village lobstermen, no cars are allowed on the island. Owned by the descendants of Thomas Edison, the island's 600 acres are preserved forever for hiking, birdwatching, beachcombing, and nature photography. Lobster Cove, on the island's south end, is easily accessible and supports the best tidepools for finding both shells and live mollusks.

Pemaquid Point, at the end of the New Harbor peninsula, features the world's most picturesque lighthouse. Striations of black and white rock form leading lines terminating at the lighthouse's base. The sight is breathtakingly spectacular. Shallow tidepools—supporting many marine denizens—are

Large tidepools harboring a variety of marine life accentuate the view of Pemaquid Point Lighthouse in New Harbor, Maine.

scattered throughout this rocky face. A strong word of caution: Always be mindful of the powerful waves that are continuously smashing against the rocks below. Should an extra-large wave cause you to slip and slide down the rocks, there is no way to get back up the rocks to safety.

An easy 45-minute drive north of Portland, **Reid State Park** has the deepest and most diverse tidepools I've ever seen. But you should be extra careful when clambering around the seaweed-encrusted rock; wear aqua shoes with good treads. When you get down into the tidepools, you'll feel like you have entered a special kingdom of the natural world. Colorful granite rocks tower above your head, obliterating the view in all directions.

Situated 25 miles due east of Portland, **Two Lights State Park** is a tidepool-lover's dream. It is

home to an immense slab of granite that extends down into the crashing surf at low tide. Large enough for a football field, this huge hunk of rock supports beautiful large tidepools. They are all relatively flat and shallow—less than a foot deep—allowing easy access for wading and investigating inhabitants. Of course, the namesake two lighthouses frame the whole scene with Downeast Maine's best lobster shack (boiled lobsters, steamers, clam chowder, corn on the cob, and blueberry cobbler—yum) located in between.

Massachusetts

Monomoy Island National Wildlife Refuge is the only true wilderness area on the Massachusetts coast. Located south of the "elbow" of Cape Cod, this magnificent mosaic of sand dunes and tidal flats is a quick ferry ride from the historic town of Chatham. The only other folks you're likely to see on the island are birdwatching groups, as Monomoy is one of the best places on the East Coast to see mixed flocks of thousands of shorebirds. Some of the more interesting shells you may find here are the calico scallop, northern quahog, transverse ark, amethyst gem clam, Gould's Pandora, lunar dove shell, solitary paper bubble, thick-lipped drill, well-ribbed dove shell, purplish tagelus, Arctic wedge clam, and knobbed whelk.

New Jersey

If you want to see a birdwatcher's eyes light with joy, just say **Cape May, New Jersey**. Located at the southern tip of the Jersey Shore, Cape May is a national historic landmark and internationally known as one of the world's top birding hot spots. During the last week of May, hordes of ancient horseshoe crabs clamber up onto Cape May's beaches and deposit billions of eggs in the sand. This smorgasbord of crab caviar attracts hundreds of thousands of migrating shorebirds, creating one of nature's greatest spectacles. While all this is happening, the tidal flats and inlets also support wonderful arrays of both living and dead mollusks. Some of the more satisfying shell finds are the ribbed mussel, channeled whelk, smooth astarte, ponderous ark, northern moon shell, sand collar, shark eye, jingle shell, Atlantic jackknife clam, eastern mud whelk, and eastern white slipper shell.

Maryland and Virginia

The **Chincoteague** and **Assateague islands** are located just off the coast of the Delmarva Peninsula, also known as the Eastern Shore in Maryland and Virginia. Both islands are known for their oyster beds and clam shoals. Known as "watermen," islanders seed and harvest their shellfish beds just as Midwestern farmers plant and harvest their corn crops. The islands receive national attention for both their immense flocks of wintering waterfowl—hundreds of thousands of snow geese, Canada geese, and trumpeter swans—and as a home to the legendary Chincoteague Island wild ponies. Descended from the livestock of seventeenth-century settlers, many of these ponies are rounded up every fall and sold on the mainland.

The best shelling is found on the southern end of **Tom's Cove Hook** in **Assateague Island National Seashore**. In particular, knobbed, lightning, and channeled whelk shells are found in abundance here. Another surprise of shelling on Assateague is the size of the oysters you will find. They tend to be much larger than oysters found on other East Coast beaches, sometimes reaching 12 inches in length.

North Carolina

Paralleling North Carolina's mainland, the Outer Banks consist of a chain of barrier islands between the open Atlantic Ocean to the east and estuarine sounds on the west. These barrier islands bear the brunt of storm surges and, in the process, protect mainland communities from serious damage. Sometimes these "ribbons of sand" are separated from the mainland by as little as one mile and as much as 30 miles. They vary in width from a few hundred yards to a mile.

Ocracoke Island covers 16 miles of a total of 180 miles of Outer Banks and provides a splendid mosaic of coastal habitats, including hammock woodlands, tidal flats, salt marshes, dune ridges, and sandy beach-

Springtime brings hundreds of thousands of shore-birds to southern New Jersey's beaches, providing an extra special experience for nature lovers of all types.

es. The National Park Service protects 80 percent of Ocracoke. A wide range of shells can be found on all of Ocracoke's beaches. Some of the more prized specimens are the Scotch bonnet, reticulated cowry helmet, lettered olive, lightning whelk, white bearded ark, crosshatched lucine, Atlantic baby's ear, ivory tusk, eared ark, disk dosinia, queen helmet, calico scallop, and turkey wing.

North Point Beach borders Hatteras Inlet and features the remains of an old coast guard station. This is the best place on Ocracoke to find olive shells, especially lettered olives. Mid-island beaches are the best locations to find Scotch bonnets, which hold the distinction of being the first state (North Carolina) seashell in the United States. The southern end of Ocracoke features many impressive shell banks, mounds of shells scattered throughout the intertidal zone. Stand and watch as waves wash over these shell piles. You'll be surprised by how many whole shells you'll find. The whelk and helmet shells in this area are especially rewarding.

Florida

If I could choose anywhere in the United States to live, I would pick **Sanibel Island, Florida**. Along with Captiva, its sister island to the north, Sanibel

Shell collecting is a wonderful family activity that truly lasts a lifetime!

has everything a naturalist and wildlife photographer could want. Its westward orientation out into the Gulf of Mexico, 23 miles from Fort Myers, makes Sanibel the perfect backstop for storm- and current-driven mollusks. In fact, Sanibel is so good at collecting mollusks that it is known as the shelling capital of North America. If you are fortunate enough to be on Sanibel during and just after a powerful winter storm, you'll quickly learn why Sanibel has this reputation. Choose any beach and you'll find yourself immersed in a seashell wonderland, including an array of 275 species of mollusks living in the shallow waters surrounding the island. Sanibel's offshore area, consisting of water 80 to 2,000 feet deep, supports another 500 molluscan species.

In addition to the spectacular shelling, Sanibel offers some of the best birdwatching in the United States, if not the world. The island's Ding Darling National Wildlife Refuge, named after a Pulitzer Prize–winning conservationist and cartoonist, annually attracts some 300 bird species.

On Florida's East Coast, **Cape Canaveral Seashore** and **Merritt Island National Wildlife Refuge** are considered bastions for both birds and mollusks. Located immediately north of the NASA Center on Cape Canaveral, these areas feature numerous small lagoons that provide critical habitat for the endangered Florida manatee, also affectionately known as the sea cow. Extra-large angel wings, up to 8 inches in length, are often found at this location.

Texas

According to the March 2007 issue of *Coastal Living Magazine*, the Texas Gulf Coast has lots of beaches well decorated by seashells. The magazine ranks **Galveston Island, Texas**, as the fourth-best shelling beach in North America, behind Sanibel Island (number one), Ocracoke Island, North Carolina (number two), and Bandon, Oregon (number three). When cold fronts roll in from the north, Galveston Island conchologists know it's time to bundle up and hit the beaches. The wind pushes the water away from shore, exposing lots of fresh tidal flats for easy pickings.

THE WEST COAST

The Pacific Coast is known for its rugged natural beauty and rich, abundant wildlife. Because of its temperate climate, the area is home to an especially rich variety of marine life, including hundreds of shellfish species. Unfortunately, West Coast beaches don't provide great bounties of shells for collectors. The often-stormy Pacific Ocean tends to pulverize shells into tiny pieces before they reach barrier islands or mainland beaches.

Despite the general lack of seashells, the Pacific Northwest is home to large, deep tidepools—incised into rocky outcrops at the base of evergreen-rimmed cliffs. Emblazoned by orange, purple, and red starfish, these mini-ecosystems are among the most colorful habitats on earth. While the brilliant colors initially divert your attention, a closer look may also reveal collectible shells or living mollusks.

From my experience, the best West Coast beaches for searching through tidepools are **Ruby Beach** in Washington's **Olympic National Park**; **Ecola State Park** near Cannon Beach, Oregon; California's **Point Lobos State Reserve** located three miles south of Carmel; and the assortment of small pocket beaches scattered along the 90 miles of the **Big Sur Highway** (U.S. Highway 1) on California's Central Coast between San Francisco and Los Angeles.

In the March 2007 issue, *Coastal Living Magazine* included two West Coast beaches in its

ABOVE
The apple murex *(Phyllonotus pomum)* is one of the stars of Gulf Coast collecting. This eye-catching beauty often can be found near the low-tide line. *Shutterstock/Bodrov Kirill Alexandrovich*

BELOW
This pond full of roseate spoonbills *(Ajaia ajaja)* feeding at sunset shows why Sanibel Island, Florida, is world-famous to birdwatchers as well as shell collectors.

listing of the top 10 beaches for shell collecting. **Bandon Oregon's** beach ranked third on the list, while **Stinson Beach, California**, took the number ten spot. When describing Bandon, the article noted that "the beaches near this charming seaside town may harbor some finds, especially in protected areas such as the mouth of the Coquille River." As for Stinson Beach, the article describes it this way: "This beach just north of San Francisco supplies limpet shells and sand dollars—plus lots of surfers, a couple of nice seafood restaurants nearby, rugged natural beauty, and endearing small-town quirkiness."

THE CARIBBEAN

Shelling in the Caribbean is by no means as successful as you might expect. Offshore barrier islands along the coast of **Belize** are moderately rich in shells. The lower Caribbean—from **Venezuela** to the islands of **Aruba**, **Bonaire**, **and Curacao**—offers a variety of cowries, spindles, and olives. Collecting on the rocky shores of the leeward sides of these islands can also yield an abundance of nerites, purpuras, pricklywinkles, and chitons.

Other Caribbean shelling locations favored by conchologists are the **Bahamas**, **St. Martin**, and the **U.S. Virgin Islands**. In the March 2007 issue, *Coastal Living Magazine* ranked **Eleuthera Island** in the Bahamas as the eighth-best shelling beach in North America. Here, the best time to go hunting for shells is after hurricanes have swept through the area, piling mounds of shells onto the beaches. Those who prefer not to wait around for a hurricane can find a huge variety of specimens by snorkeling in the shallow waters just offshore.

WHAT TO DO

Treat each shelling trip as an adventure. First and foremost, always be cognizant of your personal safety and the safety of those traveling with you. Remember that when you walk onto a beach, you are entering a

Immense sea stacks and plentiful tidepools make Oregon's Ecola State Park a can't-miss destination for seashell collectors of all ages.

natural ecosystem where you are at the absolute mercy of the tides and the surf.

The first thing you should do after arriving at your destination is to check the local tide charts. (Tides can vary widely from beach to beach, so make sure you're specific about where you're going!) Tide charts are always available in the local newspaper, at tackle shops, cafés, and diners, or from the local radio or television stations. Remember that each tidal cycle is six hours. The best shelling occurs on a receding tide, so plan on arriving about three hours before low tide. This will allow you to collect for five to six hours, through low tide to mid-tide on the next rising tide cycle. Paying attention to the tide is essential to your collecting success, as well as your health and safety. The last thing you want to do is to get washed away by an unexpectedly large wave or trapped on a sandbar by a rising tide.

After you hit the beach, always use extreme caution walking around inside tidal flats, wide inlet streams, or seaweed-covered rocks. Without a doubt, the slickest surfaces I've ever been on were the wet, slimy-green granite rocks surrounding Maine tidepools. One false step and I would have been shark bait for sure! As a general rule of thumb, never walk on green rocks.

The success of shell collecting hinges on, believe it or not, the words of Yogi Berra: "You see a lot just by looking." The power of tightly focused observation allows you to see things that other people completely miss. Serious shell collectors are a lot like police detectives, always looking for clues that will lead them to their quarries. This is also where a keen knowledge of each shell comes into play. As you walk along, you should concentrate on the key features of the shells you are seeking. For example, if you're looking for an olive shell, think about what it looks like. Picture it rolling up onto a beach and then quickly rolling back again as a wave retreats. This will allow you to recognize an olive shell instantly, even when it only partially appears in the surf. Being patient and looking closely for the shells you seek, in a location where you know they occur, will allow you to find them.

California's 1,000-plus miles of coastline ensure that there will be something—such as Pigeon Point Lighthouse—for everyone along the way.

When walking a beach, keep in mind the three tidal boundaries. The high-tide line—also known as the wrack line because that's where everything gets deposited when the tide recedes—is an excellent place to look for empty shells. Always make sure to look under seaweed, driftwood, and anything else that may be shielding a perfect shell from your view.

The intertidal zone is the area between each day's high and low tides. Here, you'll find both empty shells and living mollusks. Always be on the lookout for signs, such as bivalve siphon holes in the sand, that indicate a mollusk is down below. Finally, the low-tide line demarcates the start of the beach that is always underwater. The force of the breaking waves creates a longitudinal ditch just below the low-tide

Contrary to what you might expect, the Caribbean Islands have very few good shelling beaches. *Shutterstock*

line. The ditch is usually a few inches deep at low tide and a few feet deep at high tide. Here, the majority of your finds will be living mollusks.

Sandbars—offshore ridges of sand created by varying currents—also offer excellent shelling opportunities. Just wade out to the sandbar at low tide and see what you can find. When doing this, always use extra caution so that you don't get caught in a rip current or trapped by an incoming tide. If you do happen to get caught in a rip current, always swim parallel to the shore until you get out of it. Even the strongest swimmers will wear themselves out trying to swim against the power of a rip current.

Another collecting method for strong, experienced swimmers is snorkeling. Wearing a face mask with a breathing tube allows you to more clearly see live shells as they move around the ocean floor. Two of my greatest experiences as a naturalist were snorkeling through the crystal-clear waters of Bermuda and the Galapagos Islands. The colorful diversity of tropical fish, living shells, and crustaceans was breathtaking. Swimming with white-sided dolphins, green sea turtles, and sea lions added an extra-special touch.

The success of any shell-collecting venture always depends on the weather to a certain extent. Ironic as it may sound for a trip to the beach, you want it to rain . . . and rain hard! The stronger the storm, the more mollusks will be ripped from their ocean beds and burrows and washed ashore for you to find.

CARING FOR YOUR COLLECTION

Caring for a shell begins when you pick it up. Proper handling is required to avoid damaging each shell. This is especially true for the extra-fragile species, such as angel wings and jingle shells, as well as the murexes, crown conchs, and other spiny species. One careless move could clip off a spine or snap off an edge, spoiling the quality of what had been a perfect shell.

Now here's where you use all that special equipment you packed before you left home. Place your medium-size shells in the one-gallon plastic bags, and the small shells in the sandwich-size bags. Put the large durable shells in your backpack and the fragile shells in the hard plastic containers. Pack all shells according to size and weight, so that the heavier shells don't crush the smaller or more fragile shells.

When you get back to your motel room, unpack the shells, let them dry, and then repack them for the trip home. When repacking, wrap the fragile shells in paper towels and then layer them in the hard plastic containers.

After you get home, clean your new collection as soon as you can. Boiling is the fastest—and safest—

way to accomplish this. The length of time you boil each shell depends on its size. Boil smaller shells for one to three minutes and larger shells for five to eight minutes. Always watch the clock while you're doing this, as boiling shells for too long removes their natural oils and destroys their luster.

First, place the shells in room-temperature water, then gradually bring the water to a boil. After boiling the shells, let them cool. Next, use a pair of tweezers or a nut pick to carefully remove any parts of the animal that remain inside each shell. Always pull gently and slowly so that the animal does not break off inside the shell. After the animal is removed, soak the shell in a solution of bleach (30 percent) and water (70 percent) to remove any remaining organic material. After the cleaning process is complete, rinse the shells with clean water. If barnacles, coral, or other attached shells need to be removed, carefully scrape them off the primary shell with a small knife or nut pick.

To complete the process, rub mineral oil on the outsides of the shells to restore their luster. Never apply lacquer to your shells if you want them to keep their natural quality.

No matter what you find, spending a sunny, breezy day walking a remote beach is always one of life's best experiences! *Shutterstock/Mitch Aunger*

Afterword

As you've just read, the mollusks that create seashells are wonderfully diverse, forever fascinating animals. But mollusks are only one thread of the gigantic biological quilt that covers this planet. From tiny scarab beetles to leviathan sperm whales, the earth's ecosystems are full of creatures so wondrous and intriguing that they each merit books of their own. From your own neighborhood to the grand savannahs of equatorial Africa and the precipitous peaks of the high Himalayas, each step you take in the natural world is a moment to behold and savor.

Since you're reading this book, I suspect that you—like me—are concerned about how to protect the phenomenal natural legacy that is on this earth. I believe the solution lies in one word: education. We must all do what we can to increase young people's awareness of and appreciation for the wonders of nature.

You can do this in a number of ways: organize a neighborhood camping trip to a state or national park; plan a field trip to a science museum, wildlife sanctuary, or land trust property in your community; or conduct a group outing to a piece of land that deserves preservation.

Anything we can do to get the younger generation away from their computer screens and video games and into the great outdoors is a plus. Let them see exactly why you're so excited about seashells and the other natural wonders that surround all of us. Then they'll pass the same message along to their children and we'll all keep the earth alive and well—one generational thread at a time.

If you're concerned about the phenomenal natural legacy that is on this earth, the best way to protect it is to educate each younger generation about all of this planet's greatest wonders, including seashells.
Shutterstock/Irineos Maliaris

References

Abbott, R. Tucker. *Collectible Florida Shells.* St. Petersburg, FL: Great Outdoors Publishing Company, Inc., 1984.

Abbott, R. Tucker. *Kingdom of the Seashell.* Melbourne, FL: American Malacologists, 1972.

Dance, S. Peter. *Out of My Shell.* Sanibel Island, FL: C-Shells-3, Inc., 2005.

Dance, S. Peter. *Smithsonian Handbooks, Shells.* New York, NY: Dorling Kindersley, 1992.

Harbo, Rick M. *Shells and Shellfish of the Pacific Northwest.* Madeira Park, British Columbia: Harbour Publishing, 1997.

Lindbergh, Anne Morrow. *A Gift from the Sea.* New York, NY: Pantheon Books, 1983.

National Audubon Society. *Field Guide to Shells.* New York. NY: Chanticleer Press, Inc., 1981.

Robinson, Chuck and Debbie. *The Art of Shelling.* Manasquan, NJ: Old Squan Village Publishing, 1995.

Williams, Winston. *Florida's Fabulous Seashells and Seashore Life.* Hawaiian Gardens, CA: World Publications, 2005.

Wittkopf, Harlan E. *The Sanibel Kaleidoscope.* Algona, Iowa: Shell Island Resources, Inc., 1997.

Index